THE GENESIS OF TECHNOSCIENTIFIC REVOLUTIONS

The Genesis of Technoscientific Revolutions

RETHINKING THE NATURE AND NURTURE

OF RESEARCH

VENKATESH NARAYANAMURTI

and JEFFREY Y. TSAO

HARVARD UNIVERSITY PRESS

Cambridge, Massachusetts & London, England

2021

First printing

Library of Congress Cataloging-in-Publication Data

Names: Narayanamurti, Venkatesh, 1939– author. | Tsao, Jeffrey Y., author.

Title: The genesis of technoscientific revolutions : rethinking the nature
  and nurture of research / Venkatesh Narayanamurti and Jeffrey Y. Tsao.

Description: Cambridge, Massachusetts : Harvard University Press, 2021. |
  Includes bibliographical references and index.

Identifiers: LCCN 2021012216 | ISBN 9780674251854 (cloth)

Subjects: LCSH: Research—Management. | Research—Methodology. |
  Technology—Research—Methodology.

Classification: LCC Q180.55.M3 N37 2021 | DDC 001.4—dc23

LC record available at https://lccn.loc.gov/2021012216

To our parents,
with love, gratitude, and profound respect

CONTENTS

Introduction

Science and technology have been key to the success and well-being of humanity, and will almost certainly continue to be so. In just this past century and a half alone, our daily lives have been revolutionized by scientific advances such as the theory of special relativity and the transistor effect, and by technological advances such as the light bulb, the transistor, the laser, the blue LED, and the iPhone. One can only imagine what advances in the coming century might prove likewise revolutionary—ultrapowerful quantum information systems, general artificial intelligence, productive and human-centered global socioeconomic systems, sustainable life on earth and in space—not to mention advances that we cannot even imagine but will likely prove even more profound. As eloquently articulated by US president Jimmy Carter

(Carter, 1981) and highlighted in a recent influential policy report (American Academy of Arts and Sciences, 2014, p. 27):

> Science and technology contribute immeasurably to the lives of all Americans. Our high standard of living is largely the product of the technology that surrounds us in the home or factory. Our good health is due in large part to our ever-increasing scientific understanding. Our national security is assured by the application of technology. And our environment is protected by the use of science and technology. Indeed, our vision of the future is often largely defined by the bounty that we anticipate science and technology will bring.

Because science and technology are central to modern life, public support for advancing them—in the formal process society knows as research and development (R&D)—is substantial. But substantial support of R&D means little unless the R&D that is supported is *effective*. We are particularly concerned about research, the precious front end of R&D and the genesis of technoscientific revolutions that change the way we think and do. While development and research are both vital, research is far more fragile. Research is a deeply human endeavor and must be nurtured to achieve its full potential. As with tending a garden, care must be taken to organize, plant, feed, and weed—and the manner in which this nurturing is done must be aligned with the nature of what is being nurtured.

From our vantage point as practitioners of research, however, we have witnessed the emergence of three widespread yet mistaken beliefs about the nature of research—beliefs that are misaligned with its effective nurturing (Narayanamurti & Tsao, 2018).

The first widespread yet mistaken belief is that technology is subservient to and follows from science and, thus, that the advance of science (so-called basic research) is the pacesetter of the advance of technology (so-called applied research). This belief, stemming in part from Vannevar Bush's seminal report "Science, the Endless Frontier" (Bush, 1945), is limiting because it conflates research with science, hence narrowly confines research to the creation of new science and explicitly *not* to the creation of new technology. In fact, scientific and engineering research feed off of each other, mutually advancing in cycles of invention and discovery

(Narayanamurti & Odumosu, 2016), as exemplified by the deeply inter-active and virtually simultaneous engineering invention of the transis-tor and scientific discovery of the transistor effect at the iconic Bell Labs in 1947 (refer to the case study in Chapter 1). To emphasize the impor-tance of the symbiotic union between science and technology, we will in this book call that union *technoscience*.

The second widespread yet mistaken belief is that the goal of re-search is to answer questions. This belief is limiting because it misses the complementary and equally important finding of new questions. In Al-bert Einstein's words (Einstein & Infeld, 1971, p. 92):

> The formulation of a problem is often more essential than its solution, which may be merely a matter of mathematical or ex-perimental skill. To raise new questions, new possibilities, to re-gard old questions from a new angle, requires creative imagina-tion and marks real advance in science.

Finding a new hypothesis (a new question) is just as important as testing that hypothesis (answering that question), but is far less sup-ported in today's research environments. If Albert Einstein were now to propose research into the relationships between space, time, mass, and gravity, he would have difficulty getting funded; but Arthur Eddington, who tested Einstein's theory of general relativity, wouldn't. Charles Dar-win, who came up with the theory of evolution by natural selection, would have difficulty getting his research funded, but a test of Darwin's theory wouldn't. In fact, *both* question-finding and answer-finding are vi-tal to research and bolster each other in a symbiotic union.

The third widespread but mistaken belief stems from the "Wall Street" perspective that gained strength in the latter half of the twenti-eth century: the primacy of short-term and private return on invested capital. This belief, when applied to research, is limiting because it blinds us to the value of long-term and public return on invested capital. Truly path-breaking research seeks surprise. It overturns previous ways of do-ing and thinking in ways that cannot be anticipated—both in terms of *when* they will occur and *whom* they will benefit. Much of the benefit of research is long-term and public (extending beyond the organization that performed the research) rather than short-term and private (con-fined to the organization that performed the research). This has been true even for *private* industrial research laboratories, including the iconic ones active in the twentieth century, such as Bell Labs, IBM, Xerox PARC, Dupont, and GE. These laboratories shared common traits, such as research cultures that emphasized learning and surprise, and an irreverence for boundaries of all kinds—between disciplines, between science and technology, and between finding questions and finding answers. As a consequence, their contributions had enormous long-term public benefit. Examples of their scientific contributions include information theory, the 2.7K cosmic microwave background, electron diffraction, scanning tunneling microscopy, high-temperature superconductivity, laser-atom cooling, and fractional quantization of electronic charge. Examples of their technological contributions include the transistor, the semiconductor laser, solar cells, charge-coupled devices, the UNIX operating system and C programming, the ethernet, the computer mouse, polymer chemistry, and synthetic rubber. When, instead, short-term private benefit crowds out long-term public benefit, R&D becomes weighted away from research, whose outcomes are less certain, toward development, whose outcomes are more certain. Ultimately, such a shift in the 1980s and 1990s caused the demise of research at the great industrial research laboratories.

In this book, we present a modern rethinking of the nature and nurturing of research, with the aim of significantly improving the effectiveness of research. By considering the nature and nurturing of research as an integrated whole, we focus on those aspects of the nature of research most germane to its effective nurturing, and likewise we focus on those aspects of the nurturing of research necessary for alignment with its nature. Going forward, our hope is that the nature and nurturing of research become a powerful positive feedback loop, as illustrated in Figure 0-1, in which society continues to better understand the nature of research so as to improve its nurturing, all the while experimenting with its nurturing so as to inform and improve an understanding of its nature (Odumosu, Tsao, & Narayanamurti, 2015). To that end, our hope is also that the audience for this book will be broad and include both those interested in the nature and "understanding" of research and those interested in the nurturing and "doing" of research.

Our rethinking of the nature of research, illustrated at the top of Figure 0-1, is organized around correcting the three widespread yet

The Punctuated Equilibria of Knowledge Evolution:
Surprise and Consolidation

NATURE OF RESEARCH

The Technoscientific Method: Science, Technology, and Their Coevolution

The Intricate Dance of Question-and-Answer Finding

Nurture People with Care and Accountability

Embrace a Culture of Holistic Technoscientific Exploration

NURTURING OF RESEARCH

Align Organization, Funding, and Governance for Research

FIGURE 0-1. The feedback loop between the nature and nurturing of research, showing both the stylized facts associated with the nature of research (*top*) and the guiding principles associated with the nurture of research (*bottom*).

mistaken beliefs discussed above. Borrowing a phrase from the social sciences, we refer to the three corrected versions as "stylized facts" because they are empirical observations about the nature of research that we believe are true, general, and important. First, science and technology coevolve interactively to create new science and technology in what might be called a larger technoscientific method. Second, technoscientific knowledge is organized into seamless webs of question-and-answer pairs, and finding new questions and answers are *both* essential parts of an intricate dance that creates new question-and-answer pairs. Third, technoscientific knowledge evolves by the extension and consolidation of conventional wisdom, punctuated occasionally by surprise—and such surprise and the eventual impact of that surprise cannot be predicted or anticipated within conventional wisdom.

Importantly, our rethinking of the nature of research is both reductionist and integrative. On the one hand, our rethinking breaks technoscience and its advance into fundamental categories and mechanisms: science and technology, questions and answers, and surprise (which we identify with and research) and consolidation (which we identify with development). On the other hand, our rethinking empha-sizes powerful feedbacks between the categories and mechanisms, with technoscience and its overarching advance a unified whole much greater than the sum of its parts. Moreover, although our rethinking of the na-ture of research was, in large part, motivated by a desire to better under-stand how to nurture research, we are cautiously optimistic that it will also be of value beyond that immediate motivation.

Our rethinking of the *nurturing* of research, illustrated at the bottom of Figure 0-1, is organized around three guiding principles that center on, respectively: aligning organization, funding, and governance for re-search; embracing a culture of holistic technoscientific exploration; and nurturing people with care and accountability. In developing these guid-ing principles, we drew lessons from two sources. First, we drew from our own and others' experiences on what it means to "do" research. Our own experiences in research practice and the experiences of research leaders who nurtured spectacularly effective research organizations, including the iconic Bell Laboratories, comprise our "data." Second, we drew from our rethinking of the nature of research: how best to align the nurturing of research so that the various mechanisms associated with the nature of research are healthy, and the feedback loops and internal amplifications between them are not short-circuited.

Throughout, we benefitted from the perspectives of distinguished scholars of research: Thomas Kuhn and Brian Arthur from the history and philosophy of science and technology; Stephen Jay Gould, Herbert Simon, Philip Anderson, Stuart Kauffman, and Joseph Schumpeter from the evolutionary biological, complexity, physical, and economic sciences; and Ralph Bown, Vannevar Bush, and Donald Stokes from the world of research leadership and policy. But, drawing on our experiences within technoscientific research practice, we reframe those perspectives in language that can be followed not only by scholars, but also by practi-tioners, of research.

This book is not a casual read, but we hope it will be a rewarding one. It contains a significant rethinking of the nature and nurturing of research—with new ideas as well as old ideas integrated in new ways. To help you navigate these ideas, and to serve as a useful summary, we offer the following roadmap to Chapters 1–3 on the nature of research and Chapter 4 on the nurturing of research.

The Technoscientific Method: Science, Technology, and Their Coevolution

In Chapter 1, we discuss our first stylized fact about the nature of research: that science and technology are two qualitatively different categories of human technical knowledge, and that they coevolve deeply and interactively to create new science and technology. For *science*, we will use as our working definition facts and their explanations. By "facts," we mean abstract generalizations of empirical observations that are accepted as conventional wisdom. By "explanations," we mean, at the shallowest level, explanations of those facts; or, at deeper levels, explanations of shallower explanations. For *technology*, we will use as our working definition human-desired functions and the forms that fulfill those functions. By "forms," we mean both artifacts and processes: artifacts as the "hardware" by which we manipulate and observe the world and processes as the "software" by which artifacts are made and used (which can include the use of other artifacts and how *they* are made and used). We thus view science and technology as elegantly analogous: facts and explanations are to science what functions and forms are to technology.

Most importantly for understanding research, the existing bodies of science and technology coevolve interactively to create *new* science and technology in a powerful feedback loop. New technology is created by the three mechanisms of what we will call the *engineering method:* function-finding, in which functions desired by humans are found; form-finding, in which forms are found to satisfy those functions; and exapting, in which forms found for one function are co-opted to satisfy other functions. New science is created by the three mechanisms of what we will call the *scientific method:* fact-finding, in which abstract generalizations of empirical observations of interest to humans are found; explanation-finding, in which facts are given causal explanations; and generalizing, in which explanations for one set of facts are generalized to other sets of facts. We thus view the scientific and engineering methods as also elegantly analogous: function-finding, form-finding and exapting are to the engineering method what fact-finding, explanation-finding and generalizing are to the scientific method.

Especially important are the ways in which science advances technology and technology advances science. The engineering method is made much more productive when science can be used to model whether proposed forms will or will not fulfill proposed human-desired functions. The scientific method is made much more productive when technology can be used as a tool for experiments. In other words, science and technology *coevolve* in what we will call a larger *technoscientific* method: science draws on technology to discover new facts, while technology draws on science to invent new forms with which to fulfill human-desired functions. In the language of analog electronics, there is positive feedback whereby accumulated science and technology feeds back into the creation of new science and technology, resulting in "cycles of invention and discovery" (Narayanamurti & Odumosu, 2016).

The advance of science and technology is thus cyclical—neither science nor technology takes primacy. Technology is neither subservient to nor simply follows from science, and the advance of science is not the pacesetter of the advance of technology. Effective nurturing of research means nurturing whatever is ripest for advance—sometimes it is scientific advance, sometimes technological advance, sometimes both nearly simultaneously.

The Intricate Dance of Question-and-Answer Finding

In Chapter 2, we discuss our second stylized fact about the nature of research: that human technoscientific knowledge is organized into loosely hierarchically modular networks (Simon, 1962) of question-and-answer pairs and that these questions and answers evolve in an intricate dance to create new question-and-answer pairs. We will impose an arbitrary orientation to the hierarchy—questions above answers. In science, facts, which can be considered questions ("Why do the velocities of falling balls increase linearly with time?"), lie above their explanations, which can be considered answers ("Because uniform acceleration causes linear increases in velocity"). Those explanations can in turn be considered newly revealed questions ("Why is acceleration uniform?") that lie above deeper explanations, which can be considered deeper answers ("Because gravity, the force that causes acceleration, is uniform"). In technology, an iPhone, considered as a "how to produce certain functions" question, lies above its constituent modules (multitouch display, integrated circuit chips, camera and flash), which, taken together, can be considered a collective answer to that question. Each of those constituent modules can be considered a newly revealed functional question ("How do we produce a multitouch display?") that lies above deeper constituent sub-modules (Gorilla Glass®, capacitive multitouch surfaces) that, in turn, can be considered answers to those functional questions.

There are both similarities and differences in knowledge as one goes "up" or "down" the hierarchy. Knowledge is similar in that, at every level of the hierarchy, concepts and entities are equally challenging to explain and create: upper levels of the knowledge hierarchy (for example, biology) are no simpler or more trivial than lower levels of the knowledge hierarchy (for example, physics); they are just *different* (Anderson, 1972). Knowledge is different in that, at upper levels of the hierarchy, knowledge is more historically contingent and tacit, closer to direct human use, and easier to protect and monetize, while at lower levels of the hierarchy knowledge is more universal and formal, further from direct human use, and more difficult to protect and monetize.

Question-answer coevolution occurs in two ways: the finding of new answers and the finding of new questions. Answer-finding is the more common of the two: one begins with existing questions and seeks new answers, looking downward in the knowledge network. In science, the question of the constancy of the speed of light was newly answered by the theory of special relativity. In technology, the question of how to make an iPhone that interacts flexibly with human touch was newly answered by the multitouch display. Question-finding is the less common activity of the two, but is equally important: one begins with existing answers and seeks new questions, looking upward in the knowledge network. In science, special relativity inadvertently provided an answer to a new question of why energy is released during fission / fusion—this is the generalization mechanism in the technoscientific method mentioned above. In technology, the iPhone provided the answer to a universe of new questions not originally envisioned but later embodied in Apple's App Store (for example, how to conveniently interact with ride-hailing services)—this is the exaptation mechanism in the technoscientific method mentioned above.

Just as science and technology are tightly linked, question-finding and answer-finding are tightly linked. When seeking an answer to one question, an answer to another question often emerges: Louis Pasteur studied fermentation, hoping to answer the question of how to better produce wine, but he found that fermentation is a powerful tool for studying chemical transformations more broadly. Or, when trying to fit an answer to its question, a different answer to that question often emerges: John Bardeen, Walter Brattain, and William Shockley believed that a majority-carrier field-effect device would answer the question of how to create a semiconductor amplifier, only to find that a minority-carrier injection device answered it instead.

New questions and new answers are not equally easy to find. They are harder to find the further one moves from existing answers and questions—from what might be called the "possible." Nearest are question-and-answer pairs that already exist and are matched to each other in the possible, but with room for optimization and improved matching. Further are *latent* question and answers in the "adjacent possible"—where an idea-recombination step is required to transform the latent into the possible (Kauffman, 1996). Further still are questions and answers in the "*next*-adjacent possible": one combinatorial step removed from the adjacent possible and *two* combinatorial steps removed from the possible. Moreover, the further the new questions and answers are from the possible, not only is the connection more difficult to make but also the greater the unexpectedness of the connection, the greater the surprise, and the greater the potential for paradigm creation and the overturning of conventional wisdom.

Importantly, neither question-finding nor answer-finding takes primacy over the other. Effective nurturing of research means nurturing whichever avenue is ripest for advance—sometimes it is question-finding, sometimes it is answer-finding, sometimes it is both nearly simultaneously.

The Punctuated Equilibria of Knowledge Evolution: Surprise and Consolidation

In Chapter 3, we discuss our third stylized fact: that the evolution of knowledge occurs via punctuated equilibria, in which periods of steady and relatively more gradual advance and improvement ("consolidation") are punctuated by surprising and relatively more sudden advances ("surprise").

Paradigms mediate punctuated equilibria in much the same way that species mediate punctuated equilibria in biological evolution (Gould &

Eldredge, 1993). Paradigms are holistic combinations of knowledge "put to work" to advance technoscientific knowledge—a kind of metaknowledge of how to *use* knowledge to accomplish dynamic *changes* to knowledge. The creation of a new paradigm represents a relatively sudden and surprising break from conventional wisdom, analogous to "creative destruction" in economics (Schumpeter, 1942). The extension of an existing paradigm represents a relatively gradual consolidation and strengthening of conventional wisdom.

Paradigm creation and extension mediate the natural rhythm of surprise and consolidation, each naturally spawning the other. The creation of a new paradigm opens up a sort of "open space" for the extension of the paradigm. The extension of a paradigm in turn sows the seeds for the creation of yet newer paradigms. As a paradigm strengthens knowledge, knowledge advances. These advances may be individually small, but they accumulate, ultimately crossing performance thresholds that enable completely new questions to be asked and answered: the increasing power of integrated circuits and the increasing availability of data crossing a threshold suddenly enables computational image classification with superhuman accuracy, or the increasing power of particle accelerators crossing the energy threshold for measurable production of the Higgs boson. As a paradigm saturates in performance (on occasion, even reaching a dead end), it can catalyze new paradigms to take its place or allow new subparadigms to take the place of subparadigms on which it relies: an accumulation of seemingly unexplainable facts, such as the constancy of the speed of light, catalyzing a new theory of special relativity; or a saturation in the efficiency of incandescent lighting catalyzing a new generation of ultraefficient solid-state lighting.

Importantly, paradigm creation is different from paradigm extension and should not be nurtured the same way. For paradigm creation, surprise is paramount (Tsao et al., 2019). Because surprise and its overturning of conventional wisdom by definition cannot be anticipated by conventional wisdom, paradigm creation cannot be planned like paradigm extension can. Effective nurturing of research, which seeks to surprise conventional wisdom, requires much more flexibility and exploratory freedom than the nurturing of development, which seeks to consolidate and extend conventional wisdom (Narayanamurti & Tsao, 2018).

Guiding Principles for Nurturing Research

These three stylized facts—the coevolution of science and technology, the intricate dance of question-and-answer finding, and the punctuated equilibria of surprise and consolidation—are foundational to our rethinking of the *nature* of research. In Chapter 4, we articulate a small number of guiding principles for *nurturing* research consistent with that nature. The principles are intended to be general enough to apply to a wide range of research organizations (industrial research laboratories, universities, government research laboratories). Different parent organizations may have different overarching missions (universities to educate, industrial corporations to produce and sell goods and services, government institutions to provide particular public services), and these missions may be constraining in various ways. But any organization that wishes to do research successfully must follow similar guiding principles in nurturing that research.

Our first guiding principle for nurturing research is to *align organization, funding, and governance for research.* Research should not be invested in casually—because research outcomes are highly uncertain and cannot be scheduled or determined in advance, research organizations should invest in research only if they have a purpose that can accommodate the unexpected. Research seeks to surprise and overturn conventional wisdom, while development seeks to consolidate and strengthen conventional wisdom—a deep difference in mindset that requires research to be culturally insulated (though not intellectually isolated) from development. Research must respond flexibly and opportunistically, and this requires resources to be block-funded at the organizational level to research leadership and, ultimately, to people, not projects. And research leadership is critical: research is not simply a matter of gathering researchers for "free range" pursuits; research must be orchestrated and strike a delicate balance between organizational focus and individual freedom.

Our second guiding principle for nurturing research is to *embrace a culture of holistic technoscientific exploration.* Asking people to explore the unknown means asking them to take on an uncertain, risky, and exceedingly difficult assignment, and this requires immersion in a culture exquisitely supportive of technoscientific exploration. The full technoscientific method and its science-and-technology symbiosis must be embraced; research must not be compartmentalized by whether it is inspired by curiosity or practical application; the finding of both questions *and* answers must be embraced; hypothesis-finding must be just as valued as hypothesis-testing; and the informed contrariness that facilitates overturning conventional wisdom must be embraced.

Our third guiding principle for nurturing research is to *nurture people with care and accountability.* Though not always recognized, research is a deeply human endeavor whose success at a high level requires the nurturing of people at a high level (Bown, 1953). There is the loving and empathetic care of creative and sometimes idiosyncratic (even contrarian) researchers who are humans first, intellects second. There is also the selectivity, fairness, and accountability associated with aspiring to the highest standards of excellence in research outcomes.

What's Next

We invite you to embark with us on this journey into our rethinking of the nature and nurture of research. We have written this book so the journey can be taken from beginning to end, but the journey certainly does not need to unfold in a linear manner. Those more interested in the nurturing of research may want to skip to Chapter 4 (using the glossary for help), before reaching back to the relevant parts of Chapters 1–3 to clarify particular aspects of the nature of research; those more interested in the nature of research may want to dive into Chapters 1–3 and save Chapter 4 for the end. Those more interested in how the abstract concepts discussed in the book map to examples of particular technoscientific advances are encouraged to dive into the penultimate sections of Chapters 1–3, where we summarize relevant aspects of the histories of the transistor, the maser / laser, special relativity, the iPhone, and artificial lighting; but those who are more interested in the conceptual aspects should feel free to skip these sections.

Brief and lengthy examples have been sprinkled liberally throughout the book, in the hope that they will help concretize the many new and abstract concepts contained in the book. However, we are aware that the examples will be more easily followed by readers with certain technical backgrounds. Our examples with respect to the nature of research are drawn largely from the physical sciences and engineering, not because equally compelling examples couldn't be drawn from the life sciences and medicine, or from the information sciences and software, but because of our own deeper and idiosyncratic backgrounds in the former. With respect to the nurturing of research, we draw examples largely from US research organizations, not because equally compelling examples couldn't be drawn from international research organizations, but because of our own deeper and idiosyncratic experience with the former.

Whatever order your journey, we are confident you will find some familiar perspectives and be challenged by other, less familiar, perspectives. We hope you will find the unfamiliar perspectives reasonable, but, if you do not, we welcome the dissonance as an opportunity for our "nature and nurturing of research" community to "learn to learn." Though carefully considered, our perspectives are necessarily interim on this complex topic we believe to be vital to the future of humanity.

1

The Technoscientific Method

Science, Technology, and Their Coevolution

We discuss, in this chapter, our first stylized fact associated with the nature of research: that science and technology are two equally important but qualitatively different repositories of human knowledge and that these repositories coevolve deeply interactively. This fact is obvious to bench researchers: thinking (trying to understand how things work) and doing (trying to build and operate things that work) go hand in hand. As discussed in the introduction, however, an opposing belief is widespread but mistaken: the belief that technology is subservient to and follows from science and, thus, that the advance of science is the pacesetter of the advance of technology. In this chapter, we correct that belief.

We begin with basic definitions and characteristics of science and technology along with their relationship to human culture. Science, technology, and culture, as illustrated in Figure 1-1, we denote by the symbols S, T, and C.

We then discuss the mechanisms by which science and technology coevolve and grow. The growth of S and T we denote by $\dot{S}$ and $\dot{T}$, adopting the mathematics notation for dots above quantities to denote time rates of change of those quantities. The notation is symbolic rather than quantitatively definitional—the dynamic evolution of S and T could not possibly be represented as rates of change of single scalar quantities—but we believe it is important. It is common practice to use the words "science" and "technology" ambiguously to refer sometimes to the static repositories themselves (the "making use" of science and technology) and sometimes to the dynamic growth of those repositories (the "doing" of

CULTURE (C)

SCIENCE (S) TECHNOLOGY (T)

FIGURE 1-1. Three categories of human knowledge represented by Albert Einstein, iconic discoverer of scientific explanation; Thomas Edison, iconic inventor of technological functions and forms; and human culture's desire to migrate and explore. *Credit:* Photo of Thomas Edison from Record Group 306, Records of the U.S. Information Agency (306-NT-279T-20), National Archives and Records Administration. Photo of Albert Einstein from the Österriechische Nationalbibliothek. Photo of wagon train from the Library of Congress, Prints and Photographs Division, LC-DIG-ppmsca-33293.

science and technology, which, for technology, we identify with engineering). The theory of special relativity as a static piece of scientific knowledge (S), for example, might be used to dynamically engineer a new global positioning system technology ($\dot{T}$), but that is not the same as dynamically extending or perhaps even overturning the theory of special relativity itself ($\dot{S}$). An optical telescope as a static piece of technological knowledge (T) might be used to establish new scientific facts about planetary motion ($\dot{S}$), but that is not the same as dynamically engineering a new telescope or perhaps even replacing it with radio astronomy ($\dot{T}$). Our focus is on the growth of science and technology, hence on $\dot{S}$ and $\dot{T}$, but because the growth of science and technology "stands on the shoulders" (to borrow the words of Isaac Newton) of existing science and technology, we will also spend some time discussing the nature of S and T themselves.

Finally, we discuss the interactive coevolution of science and technology using as examples two iconic technoscientific advances of the twentieth century: the transistor and the maser / laser. Both advances were transformative—the transistor primarily through its impact on computation, the maser / laser primarily through its impact on communication.

1.1 The Repositories of Technoscientific Knowledge: *S* and *T*

The two repositories of human knowledge of primary concern to us are scientific and technological knowledge—what we will often simply call science and technology. To these we add, for completeness, a third repository of knowledge: human culture. Culture is mediated by, and in turn mediates, the growth of science and technology—so all three coevolve. These three repositories of knowledge are illustrated in Figure 1-1, but before we describe them in more detail, we make a few overarching comments.

First, we define science and technology very broadly. By "science," we mean humanity's understanding of the world, where "the world" is all-encompassing. Science thus includes humanity's understanding of the stars in the galaxy (the natural physical world); the plant, animal, and human biosphere on earth (the natural living world); the technology humanity has created (the human-synthesized world); and human society itself (the human social world). By "technology," we mean humanity's means for interacting with the world. Technology thus includes humanity's use of electromagnetics (natural physical phenomena), animals or plants (natural living phenomena), and augmentations of our own eyes and hands (augmented human biological phenomena).

Second, our distinctions between science, technology, and culture are functional, mapping to functions that any natural or artificial intelligence adapting to an environment must perform. Any such intelligence must first *interact* with the world by sensing and actuating, which it would do via technology. Second, it must *predict* how its interactions with the world lead to various outcomes, which it would do via science. And, third, it must *know* which outcomes to prefer; it must have an "objective function" that scores various outcomes—an objective function that culture provides. Roughly speaking, technology maps to "know how" knowledge, science maps to "know why" knowledge, and culture maps to "know what" knowledge (Garud, 1997).

Third, we do not distinguish between science and technology based on how they are culturally valued, though how they are valued is certainly important for the nurturing of research. Technology is often thought of as more valuable financially than science because its transfer can be more easily protected and monetized, while science is often thought of as less valuable financially because its transfer is less easily protected and monetized. Thus, the rewards for new technology are more often financial and the environments within which new technology is created more often secret, while the rewards for new science are more often reputational and the environments within which new science is created more often open (Dasgupta & David, 1994). We do not regard these distinctions as fundamental, however. One can easily imagine a cultural value system in which financial rewards for new technology are eschewed: Ben Franklin did so with his inventions, famously saying that "As we enjoy great advantages from the inventions of others, we should be glad of an opportunity to serve others by any invention of ours, and this we should do freely and generously" (Franklin, 1791, p. 216); and AT&T Bell Laboratories did so with its invention of the transistor and the Unix operating system, which it patented but then licensed nearly free of cost (Brinkman et al., 1997). One can also easily imagine a cultural value system in which reputational credit drives extreme competition and secrecy (at least in the short term) in new science: Isaac Newton was famously secretive, not wanting to publish his treatise *Opticks* until his archrival Robert Hooke had passed away; likewise, the scientific knowledge gained during the Manhattan Project was understandably kept tightly secret. It is a basic human instinct to be generous and share, and this instinct can be applied to both engineering invention and scientific discovery; it is also a basic human instinct to be competitive and seek advantage, and this instinct can also be applied to both engineering invention and scientific discovery.

Fourth, our distinction between science and technology is not based on the ease with which they can be, or on the vehicles by which they are, communicated. Science is sometimes identified with codified knowledge that can be formalized and transmitted in writing, while technology is sometimes identified with tacit knowledge difficult to formalize and transmit in writing. Science is also sometimes identified with journal publications, while technology is sometimes identified with patents. We do not regard these distinctions as fundamental, however. There is much tacit knowledge in science and much codified knowledge in technology, and new technology is often communicated via journal publications just as it is via patents.

Science: Facts and Their Explanations

For our definition of "science," we use the following: science is humanity's library of facts about the world and explanations for those facts (Simon, 2001).

By "facts," we mean stable, observed patterns about the world. These patterns can be those actuated and sensed directly with our human biology: when one releases a ball, one can visually observe that it falls and that it falls faster the farther it has fallen. Or they can be actuated and sensed through technological instrumentation: to determine this pattern more quantitatively, one can (as Galileo Galilei did), use the technology of inclined planes to slow the fall of the ball so as to more precisely time its motion. Patterns can also have various degrees of stability. At one extreme, in the physical sciences, patterns can be extremely stable and reproducible, such as the motion of mechanical objects; at the other extreme, in the social sciences, patterns may be less stable and less reproducible, such as the behavior of human groups; and in between these two extremes lies a continuum of probabilistic certainty.

By "explanations of facts," we mean the entire hierarchy of shallow and successively deeper explanations of observed patterns about the world. Shallow explanations might be nothing more than concise rearticulations of the pattern itself: Galileo's explanation of the observed distance versus time pattern was that the velocities of falling balls increase linearly with time. A deeper explanation would be Isaac Newton's explanation that gravity is a force, that uniform forces cause uniform acceleration, and that uniform acceleration causes linear increase in velocity. Facts and explanations, as will be discussed in more detail in Chapter 2, are thus hierarchically nested: a pattern is given a shallow empirical explanation; that shallow empirical explanation is then given a deeper explanation, which may, in turn, be even more deeply explained. The deepest explanations are more powerful in that they often generalize to explanations of other patterns—Isaac Newton's explanation for patterns exhibited by falling balls also explaining patterns exhibited by planets orbiting the sun.

We call special attention to two important aspects of explanation: the more parsimonious and the more causal, the more powerful.

By "parsimonious," we mean that the explanation is simpler than the pattern it is trying to explain; indeed, that it is as simple as possible. Once a pattern is found in raw data, the data are to some extent redundant because the pattern provides a parsimonious description simpler and more concise than the data. Likewise, once an explanation is found for the pattern, the pattern is, to some extent, redundant because the explanation provides a parsimonious description simpler and more concise than the pattern. As Herbert Simon has said (Simon, 2001, p. 7):

> The primordial acts of science are to observe phenomena, to seek patterns (redundancy) in them, and to re-describe them in terms of the discovered patterns, thereby removing redundancy. The simplicity that is sought and found beautiful is the simplicity of parsimony, which rests, in turn, on the exploitation of redundancy. We do not seek the absolutely simplest law but the law that is simplest in relation to the range of phenomena it explains, that is most parsimonious.

This is just Ockham's razor, which can be thought of as a heuristic for ease of falsifiability: the simpler the explanation of a pattern (while maintaining "verisimilitude" to the pattern), the more generalizable to predict other patterns, the more easily falsifiable if the prediction does not hold, and thus if not yet falsified the more likely to be true (Popper, 2005, p. 267).

By "causal," we mean the degree to which the understanding of observed facts is not just correlational but causal. Just as with parsimoniousness, the more causally explanatory one's understanding of a pattern, the more generalizable that understanding to the prediction of other patterns. And, the more generalizable the understanding, the more easily falsifiable if the prediction does not hold, and thus if not yet falsified the more likely to be true. If the causal explanation for a ball's acceleration is an applied force, perhaps the general explanation for *any* object's acceleration is an applied force, which is refuted if just one object is observed to accelerate in the absence of a force. We know we cannot prove the world to be causal, but, as a practical matter, causality clearly "works" and has been deeply imprinted in the way humans explain observed patterns and structure their world model. As articulated by Judea Pearl (winner of the 2012 Turing Award) and Dana Mackenzie (Pearl & Mackenzie, 2018, p. 24):

> Very early in our evolution, we humans realized that the world is not made up only of dry facts (what we might call data today); rather, these facts are glued together by an intricate web of cause-effect relationships.

Causal reasoning may even be one of the missing ingredients needed for artificial intelligence to become more general and human-like.

Technology: Functions and the Forms that Fulfill Them

Just as our definition of science has two bins (humanity's library of facts about the world and the explanations for those facts); our definition of technology also has two bins: humanity's library of human-desired functions and the forms that fulfill those functions.

By "human-desired functions," we mean functions that satisfy some human desire: our desire to eat, our desire to interact with others, our desire for shelter, or our curiosity to see out into the unknown universe. We make no value judgments on those human desires. They may evolve, but at any instant in time they are what they are and may include waging war, feeding the poor, or satisfying human curiosity.

By "forms," we mean both artifacts and processes. By "artifacts," we mean hardware: devices that may be as simple as a shaped stone or as complex as an integrated circuit. By "processes," we mean sequences of operations by which those artifacts are made and used. A shaped stone might be made by chipping at a raw stone, then used for cutting by grasping and sawing. A silicon wafer might be made by slicing a silicon boule grown from a silicon melt, then used as the starting substrate for an integrated circuit. An integrated circuit, in turn, might be made by a complex set of additive and subtractive lithographic thin-film processes, then used to compute via operating system and application software.

The usefulness of technological forms (artifacts and processes) lies in their ability to "capture phenomena" to satisfy some human-desired function (Arthur, 2009). Shaped stones make use of mechanical abrasion for cutting. Silicon semiconductors make use of a set of mechanical and chemical phenomena for the processes used to make an integrated circuit. Integrated circuits make use of digital electronic storage and switching phenomena for computation. Because the worlds of phenomena and human use are both rich and complex, technological forms and functions, at the interface between the two, are also rich and complex. This richness and complexity manifests itself as a hierarchical nesting, as will be discussed in more detail in Chapter 2, similar to the nesting of facts and explanations in science. Shaped stones can be used in a plough, which in turn can be used to dig furrows, which in turn can be used to plant seeds and grow crops for human consumption. Integrated circuits can be used in printed circuit boards, which in turn can be used in computers, which in turn can be used in iPhones for direct human use. We consider all levels in the hierarchical nesting to be technologies—from the raw materials at the bottom of the hierarchy to the products with which humans interact directly.

Importantly, we view technology as neither only an extension of science nor only an application of prior scientific knowledge. Indeed, for much of human history, technological knowledge *preceded* scientific knowledge (Basalla, 1988; Layton, 1974; Rosenberg, 1982). Numerous technological inventions (like eyeglasses and the steam engine) came well before they were scientifically understood. Indeed, that technology must be able to evolve without science follows from considering technology an extension of our human biology. We used our hands to manipulate long before we used pliers; we used our eyes to see long before we used telescopes; and the evolution of neither hands nor eyes required science. As articulated by political scientist Donald Stokes (Stokes, 2011, p. 19):

> The deepest flaw in the dynamic form of the postwar paradigm is the premise that such flows as there may be between science and technology are uniformly one way, from scientific discovery to technological innovation. . . . The annals of science suggest that this premise has always been false to the history of science and technology. There was indeed a notable reverse flow, from technology to science, from the time of Bacon to the second industrial revolution.

Indeed, in that technology is phenomena captured and put to human use, it elicits phenomena from nature and is thus an enormous repository of patterns and empirical knowledge—of scientific facts to be "made sense of" by scientific explanation. Technology is a crucial interactive portal to and from the world of phenomena, a world that consistently surprises us, imaginative as we are, with phenomena that science could hardly have predicted in advance. In the words of William Shakespeare's Hamlet (I.iv.165–166):

> There are more things in heaven and earth, Horatio, than are dreamt of in your philosophy.

Culture: The Selection Pressure for $\dot{S}$ and $\dot{T}$

Our primary focus in this book is science and technology, but science and technology do not exist in a vacuum; they are embedded in human culture. Culture is, of course, complex, and it is well beyond the scope of this book and our expertise to give it a scholarly definition. But, because culture coevolves *with* science and technology, at least a working definition is unavoidable. In this book, we will think of "culture" broadly as human values, desires, interests, norms, and behaviors, and we will be concerned with those aspects that coevolve with science and technology.

How does culture coevolve with science and technology? It does so in the two directions articulated by historian Thomas Hughes (Hughes, 1987, p. 51):

> Technological systems contain messy, complex, problem-solving components. They are both socially constructed and society shaping.

In one direction, science and technology shape culture (Marx, 1994). Science and technology constantly open up new frontiers of possibilities for human values, desires, interests, norms, and behaviors: the massive urbanization evidently desired by humanity was enabled by building, transportation, and sanitation technologies, while the interconnecting of our social and physical worlds, also evidently fulfilling a human desire, was enabled by information technologies. In the other direction, culture shapes science and technology. Culture influences, as it should, our choice of *what* $\dot{S}$ and $\dot{T}$ we do. How does it do so? It does so by providing an objective function, or selection pressure, for $\dot{S}$ and $\dot{T}$. In particular, it selects for one or both of a pair of characteristics: utility (u) and learning (l). These characteristics are very different from, even orthogonal to, each other, and they will be discussed in more detail in Chapters 3 and 4. Here, we presage that discussion.

The first characteristic that culture selects for is *utility*. Our definition of utility is broad: new technoscientific knowledge has utility if it "works" and if it has an impact on human desires. The theory of special relativity "works" and has an impact on human-desired knowledge as diverse as global positioning systems and astronomy. The iPhone and the internal combustion engine both "work" and have had an enormous direct impact on how we go about our daily lives. Our definition of "human desire" is also broad: we include both practical human desires, such as to eat; and we also include the less practical human desire to satisfy human curiosity, such as discovering the origin of stars and galaxies, or inventing a laser just to see what it can do.

In the same way science and technology are nested, as discussed earlier, utility is also nested. There is "ultimate" utility, which can be tied to basic evolved human desires: the desire to explore, to survive, and to reproduce. Then there are "proximate" utilities, which can be tied to intermediate desires that make possible the ultimate desires: understanding how telescopes work so that one can explore the heavens; developing the ability to etch lithographic patterns to create power-efficient silicon integrated circuits for battery-operated smartphones so we can communicate. By "utility," we mean all these utilities, ultimate and proximate, whether of a scientific or technological nature. Moreover, we recognize that utility is not static but rather is dynamic and constantly changing. A necessity for one people, generation, or social group may have no utilitarian value or may be a superficial luxury for another people, generation, or social group. And, because the rate at which culture changes may lag the rate at which science and technology change, as a selection pressure utility may sometimes seem maladapted to human life. Utility selects for high-sugar-content foods even as in excess they lead to heart disease. Utility selects for high-violence-containing gaming when this seemingly encourages societal violence, and it selects for technologies for mass destruction when this seemingly puts all of humanity at existential risk. But all utility stems from human values, needs, interests, norms, and uses that were presumably selected for over the course of human evolution, so they at one time had fitness value to humanity.

The second characteristic that culture selects for is *learning*. The human desire to learn is basic and hardwired, evident even in infants, who are attracted to events that surprise them (Köster et al., 2019). Its origin lies in the exploration / exploitation trade-off all living things face as they adapt to new environments (Hills et al., 2015). Living things must exploit their environment in order to gain resources to sustain life in the present, but they must also explore and learn about their environment in order to better sustain life in the future. Learning can be both genetic, imprinted across generations in the DNA of organisms as they evolve, or cognitive, imprinted within a generation in the neural brain patterns of organisms. And learning applies to both science and technology—in science, we learn by finding new facts and explanations; in technology, we learn by finding new functions and forms.

We also distinguish (as we will discuss in more detail later in Chapter 3), between two kinds of learning: one that emphasizes consolidation and one that emphasizes surprise (Tsao et al., 2019). On the one hand, learning by consolidation extends and strengthens conventional wisdom. As conventional wisdom is exercised in some new way, uncertainty arises; if the conventional wisdom is found to "hold true," the uncertainty is diminished and conventional wisdom is strengthened and consolidated. On the other hand, learning by surprise overturns conventional wisdom. When we exercise conventional wisdom in some new way, it may not be confirmed but contradicted and overturned, surprised in some way. We associate learning by consolidation with the exploitation and extension of existing paradigms, while we associate learning by surprise with the exploration and creation of new paradigms.

1.2 The Growth of Technoscientific Knowledge: $\dot{S}$ and $\dot{T}$

As just discussed, science (S) and technology (T) are distinct repositories of knowledge with distinct characteristics. The science repository of knowledge consists of facts and their explanations, and the technology repository of knowledge consists of human-desired functions and the forms that fulfill those functions. Of most interest to us is not the static nature of these bodies of knowledge, but their dynamic nature—how new science and new technology are created and evolve ($\dot{S}$ and $\dot{T}$). That creation and evolution are mediated by mechanisms, illustrated in Figure 1-2, that comprise what we call the technoscientific method, the interacting combination of parallel and complementary engineering and scientific methods.

The engineering method is comprised of three mechanisms:

- Function-finding: the finding of human-desired functions to pursue
- Form-finding: the creation of forms to fulfill those functions
- Exapting: the repurposing of forms created to fulfill one function so as to fulfill another function

This engineering method encompasses what are sometimes called the "engineering sciences" and "design thinking." Engineering science emphasizes reductionist thinking, the creation of forms to fulfill functions, which we will call in Chapter 2, and hint at here, a kind of "answer-finding." Design thinking emphasizes integrative thinking, the

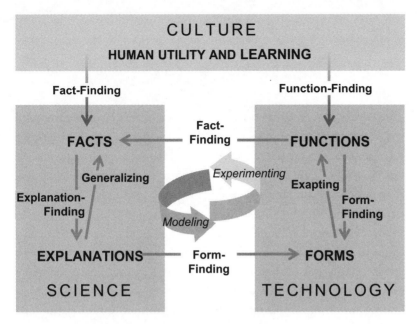

FIGURE 1-2. Mechanisms of the "technoscientific method," the engine that drives the coevolution of science and technology.

processes of finding new functions and the repurposing of forms for these new functions, which we will call in Chapter 2, and hint at here, a kind of "question-finding."

The analogous scientific method is also comprised of three mechanisms:

- Fact-finding: the finding of observed facts about nature that are of human interest
- Explanation-finding: the finding of explanations for those facts and for shallower explanations
- Generalizing: the repurposing of theory, created to explain one set of facts or shallower theory, to explain other facts or other shallower theory, including to "hypothesis-test" the theory

This scientific method maps closely to Thomas Kuhn's "three classes of problems" in normal science (Kuhn, 2012, p. 34): determination of significant fact (fact-finding), matching of facts with theory (explanation-finding), and articulation of theory (generalizing).

The engineering and scientific methods have different aims, as articulated by computer architect and Turing Award winner Fred Brooks:

> A scientist builds in order to learn; an engineer learns in order to build.

But they feed on each other in what has been called "cycles of invention and discovery" (Narayanamurti & Odumosu, 2016). They feed on each other in part through the form-finding and the function-finding mechanisms, as we will discuss these in their respective subsections below. But they also feed on each other more pervasively through experimenting and modeling.

Experimentation is inherently technological, drawing on engineered forms to fulfill the function of experimentation. Experimentation is also vital to the growth of both science and technology. Scientific experimentation is essential for finding facts, either in a less directed "we don't know what we'll find" manner, or in a more directed "we're looking for a not-yet-observed-but-predicted fact so as to test a particular explanation or theory" manner. Technological experimentation is essential for testing forms, either in a less directed "I wonder what function this form will fulfill" manner or in a more directed "does this form fulfill the function we intended it to" manner.

Modeling draws on both science and technology, drawing on scientific facts and explanations but also on calculational technologies to predict the consequences of those facts and explanations in particular situations. Modeling also feeds the growth of both science and technology. Scientific modeling is essential for connecting explanation with fact: explanations are often simple to state (planets move in response to a mass-proportional inverse-square-law gravitational force), but predicting the facts (elliptical orbits) that follow from those explanations can require sophisticated mathematical and / or calculational modeling tools (calculus). Engineering modeling is essential for connecting form with function, particularly in the design phase before prototypes have been made: forms are often simple to describe (an airplane wing of a certain shape), but understanding the functions they perform (aerodynamic lift for various wind velocities and angles) can require sophisticated mathematical and / or calculational models.

In other words, the scientific and engineering methods build on both science and technology, mediated in large part by experimenting and modeling. The more advanced the science and technology, the more advanced the experimenting and modeling. The more advanced the experimenting and modeling, the more sophisticated the scientific and engineering methods that advance science and technology. The mechanisms of the combined technoscientific method, forming as they do a cycle of invention and discovery (Narayanamurti & Odumosu, 2016), are all important and must all be nurtured. As we intimated in the introduction and will return to in Chapter 4, however, when some mechanisms are undervalued, the resulting positive feedback cycle is less than optimally nurtured.

Function-Finding

Function-finding is the first mechanism of the engineering method: the finding of human-desired function. In part, the finding of functions is determined by the inner logic of the technology in which the functions are embedded: a function with no current possibility of being fulfilled (a perpetual motion machine), no matter how desired, is less likely to be selected as something to pursue. But the finding of functions is ultimately determined by the outer logic of human culture and desires. A function that human culture deems less useful will not be selected to pursue even if it appears amenable to fulfilling. A myriad of inventions failed not because they were impossible, but because the marketplace, a proxy for human culture, did not select for them. Ford's 1958 Edsel, AT&T's 1964 Picturephone, Apple's 1993 Newton personal digital assistant—these famous product failures are just the tip of a large iceberg of technology and product failures throughout history, illustrating the primacy of the "outer logic of human culture and desires" over the "inner logic" of the technology itself.

This is not to say that human culture's ability to select what $\dot{T}$ to pursue is perfect. It is far from perfect. The contrarian thinker who goes against the conventional wisdom of prevailing culture is often the one who makes the more fruitful selections and ultimately causes human culture to change and accommodate. For example, the functional need for widespread computation was famously underestimated by Thomas Watson, president of IBM, as captured in the 1943 phrase sometimes attributed to him: "I think there is a world market for maybe five computers" (Qtd. in Rybaczyk, 2005, p. 36). Ultimately, though, it is human judgment, as misinformed or maladapted as it might sometimes be, that makes the selections for what T to pursue.

If the past is any guide to the future, functions of interest to human culture and desires, in all their heterogeneity and diversity, are seemingly infinite. They have, throughout human history, never reached, and perhaps never will reach, a saturation point. The human imagination is limitless. Once one set of functions of interest has been selected then fulfilled, a new set, not yet fulfilled, has inevitably been found. We are surrounded by latent functions (to eat more nutritiously, to communicate more seamlessly, to observe with more acuity), any of which is a potential function of interest to fulfill. Moreover, human-desired functions are numerous because they are also nested. A high-level ultimate function (to eat) spawns a number of proximate functions (to farm, to make tractors to till soil), each of which spawns a number of even more proximate functions. Indeed, it was in part his recognition of the insatiability of human-desired functions that led the economist Joseph Schumpeter in the mid-twentieth century to view entrepreneurial disruption of existing functions by new and / or augmented functions ("creative destruction"), rather than steady-state or equilibrium, as the essence of the human condition and of human economic growth (Schumpeter, 1942, p. 83).

Form-Finding

Once some human-desired function (for example, fastening two sheets of paper to each other) has been selected, we are launched into the second mechanism of the engineering method: *form-finding*: the finding of the form (for example, staples) that fulfills the human-desired function. Because of the infiniteness of human desires discussed above, however, human-desired function is never exactly defined and is often quite inexactly defined. This inexactness is the origin of the "design thinking" aspect of the overall engineering method alluded to earlier, which has an eye to clarifying or particularizing the human-desired function that might lead to better matched forms (for example, paper clips).

To find form from function, to create artifacts and processes that will fulfill a function, form-finding makes use of highly evolved vocabularies and grammars—existing stocks of components (artifacts and processes) in a knowledge domain or discipline and the rules by which these components can be appropriately combined. As articulated by Brian Arthur (Arthur, 2009, p. 76):

> A new device or method is put together from the available components—the available vocabulary—of a domain. In this sense a domain forms a language; and a new technological artifact constructed from components of the domain is an utterance in the domain's language.

Some of these vocabularies and grammars derive from science and from the facts and explanations of science.

At the level of facts, the vocabularies and grammars might draw on the properties of materials organized into so-called "engineering tables" (the heat capacities of gases at various pressures and temperatures, the stiffnesses and yield strengths of materials of different dimensions, the electrical conductivities of metals at various temperatures). Tabulations of such properties can be immensely useful with or without deeper explanations of these properties.

At the level of explanations, the vocabularies and grammars might draw on deep scientific principles, highly mathematized. The rules governing the form-finding of global positioning satellites would be impossible without knowledge of the relativistic corrections for time given by the theory of special relativity. The rules governing the form-finding of the atom bomb would be difficult without knowledge of the mass-energy equivalence relations also given by the theory of special relativity. These are situations for which the phenomena being harnessed by the forms are more esoteric, further from direct human experience (such as electromagnetic versus mechanical phenomena), or complicated by being a superposition of phenomena, each governed by different scientific explanations and laws. In these situations, the vocabularies and grammars associated with harnessing that phenomena require the deep, simplifying, and sometimes counterintuitive understanding that science provides. It also often requires augmenting scientific explanation with modeling and its tools of performance prediction: mathematics, logic, and computational simulation. The results of such predictions can often be counterintuitive because humans are limited in how many relationships they can track at once.

Scientific knowledge enables us both to design forms likely to fulfill human-desired function and, especially, to eliminate forms likely *not* to fulfill human-desired function. The space of possible forms is huge. Only a vanishingly small portion of this space represents useful forms that fulfill real human-desired functions. Finding forms can be like finding a needle in a haystack and thus can benefit from a shrinking of the haystack via scientific knowledge. Perpetual motion machines, in the light of conservation of energy, can be eliminated from consideration. Faster-than-the-speed-of-light machines, with knowledge of the theory of special relativity, can likewise be eliminated from consideration. As social psychologist Kurt Lewin once said, "there is nothing more practical than a good theory" (Lewin, 1952). Scientific knowledge is critical to identifying potential forms whose operations disobey scientific laws or theories and can thus be eliminated.

We do not want to overstate the role of science in form-finding or in the engineering method more broadly, however. Science is limited in what it can provide to engineering and the engineering method, because the regularity and generalizability that science brings is itself limited. The form-to-function connection is complex and requires hard-earned knowledge gained from experience in the field and from contact with the use environment. Though science continually expands in scope, that scope is (and perhaps will always be) a small fraction of nature. Engineering is impatient and cannot wait until the sciences have explained the universe; it must also rely on learning by doing, and it must have room for purposeful experimentation and tinkering. As will be discussed later in this chapter, one of the great inventions of the late nineteenth century, the transistor, went beyond then-current science and deep craft. As will be discussed later in Chapter 3, one of the great inventions of the late twentieth century, the blue LED by 2014 Nobel Laureates Isamu Akasaki, Hiroshi Amano, and Shuji Nakamura, even went *against* then-current scientific knowledge and deep craft (Tsao et al., 2015). These beyond-current-science new technologies serve notice that nature is consistently richer than we can imagine and the ultimate source of both new technology *and* new science. Science is a conservative force that often denies the possibility of new technologies that are inconsistent with current scientific understanding but that turn out to be possible after all.

Though there is a degree of blindness and uncertainty, and though we used above the word "tinkering," we don't mean it in the pejorative sense of "dumb luck" or to echo the "heat, beat, and hope" sentiment often heard in the corridors of academic physics departments in the 1990s with respect to exploratory synthesis of new materials. Tinkering has (and must have) an element of blindness (Campbell, 1960), but it can also be guided by intuition and hard-earned tacit knowledge, even as it does not take that knowledge as gospel. As we will discuss in Chapter 3, going beyond what we know, going contrary to conventional wisdom, is ultimately the only way to create truly new and surprising technology. However, as we will discuss in Chapter 4, doing so in an *informed* way, as an "informed contrarian," increases the odds of success.

Exapting

Having found a library of forms, we are launched into the third mechanism of the engineering method. This is "exapting," the finding of new function from existing form, the repurposing or co-opting, often opportunistic and serendipitous, of existing forms for new functions—a kind of knowledge spillover.

The conceptual origins of the term "exaptation" lie in evolutionary biology, beginning with Darwin's original term "pre-adaptation" (Darwin, 1859, pp. 179–186), then clarified and renamed "exaptation" by evolutionary biologists Stephen Jay Gould and Elisabeth Vrba (Gould & Vrba, 1982). A classic example is dinosaur feathers, likely adapted originally for the purpose of thermoregulation but then exapted (co-opted) for flight, ultimately leading to the magnificent flying abilities of the modern bird (Gould & Vrba, 1982). Another example is the sutures in vertebrate skulls, likely adapted originally to enable postbirth growth but then exapted (co-opted) to enable the large heads of mammals to be molded as they passed through narrow birth canals at birth (Gould, 1991).

More recently, exaptation as a concept and term was itself co-opted and applied to technology (Dew et al., 2004; Kauffman, 2019). The transistor, invented to replace vacuum tubes for communications, was enhanced and utilized in ways not foreseen or even imagined by the inventors—immediately after its invention in 1947, it was co-opted and combined with other miniaturization technologies in the 1950s by the Japanese company Sony to produce the first ultracompact portable radios. Similarly, the laser, originally invented both out of sheer curiosity and to extend the field of microwave spectroscopy by making electromagnetic wave oscillators at frequencies higher into the infrared, was combined with fiber optics to revolutionize long-haul telecommunications. Magnetrons, invented for emission sources for radar waves during World War II, were repurposed to create microwave ovens. Drugs developed to relieve the symptoms of one disease are often repurposed to alleviate the symptoms of another disease. Exaptation, we now understand, is ubiquitous and a key mechanism in the overarching technoscientific method.

Though exaptation benefits from luck and serendipity, it is not *just* luck and serendipity. It requires a preparedness. As famously noted by Louis Pasteur, "chance favors only the prepared mind" (Pasteur, 1854). It requires a mind that is simultaneously alert to forms that have been created for some function as well as to a stock of latent functions waiting for forms to fulfill them. The British surgeon Joseph Lister exapted carbolic acid, previously used for preventing odors from emanating from refuse material, for the unexpected purpose of postsurgical antisepsis (Andriani & Cattani, 2016). Lister was simultaneously alert to an existing artifact (carbolic acid) and the function for which it was being used, as well as to a latent function that he knew was unfulfilled but for which the artifact might be useful.

Interestingly, the *results* of both form-finding and exapting are the same: a new matching between a human-desired function and a form that fulfills that function. The difference is one of perspective. If one's perspective is that of a known human-desired function, then one seeks a form, any form, that fulfills that function; if one's perspective is that of a known form, then one seeks a human-desired function, any function, that is fulfilled by that form.

Fact-Finding

Fact-finding is the first mechanism of the scientific method: the finding of stable observed patterns about the world. It is both precursor to explanation-finding, the explaining by theory of facts just found, and the necessary follow-up to explanation-finding, the testing of theory by its prediction of new facts to be found.

Just as for function-finding in the engineering method, the selection of what facts to pursue in the scientific method is in part determined by the inner logic of the technoscience in which those potential facts are embedded. A potential fact with no possibility of being found, no matter how interesting, is not likely to be selected to pursue. But the selection is also in part determined by the outer logic of human culture: a fact that human culture deems uninteresting will have difficulty being selected to pursue *even if it is where the most can be learned*. As articulated by sociologist Robert Merton (Merton, 1973, p. 59):

> To some of his contemporaries, Galileo and his successors were obviously engaged in a trivial pastime, as they watched balls roll-ing down inclined planes rather than attending to such really important topics as means of improving ship construction that would enlarge commerce and naval might. At about the same time, the Dutch microscopist, Swammerdam, was the butt of ridicule by those far-seeing critics who knew that sustained at-tention to his "tiny animals," the micro-organisms, was an un-imaginative focus on patently trivial minutiae. These critics of-ten had authoritative social support. Charles II, for example, could join in the grand joke about the absurdity of trying to "weight the ayre," as he learned of the fundamental work on at-mospheric pressure which to his mind was nothing more than childish diversion and idle amusement when compared with the Big Topics to which natural philosophers should attend. The history of science provides a long if not endless list of instances of the easy confusion between the seemingly self-evident trivial-ity of the object under scrutiny and the cognitive significance of the investigation.

Thus, it is important for a researcher sometimes to choose to study facts that current culture and conventional wisdom believes uninteresting, if the researcher has reason to believe they are in fact profound. Doing so, however, requires an "informed contrarian" kind of attitude, one that we will discuss in Chapter 4.

Because technology (and its biological antecedents) provides the means by which we interact with the world, technology is at the heart of fact-finding. It is at the heart in two distinct and powerful ways: as a *tool* for observation and as an *object* of observation. As a tool for observation, technology reveals extrinsic phenomena that lie outside of the technology being used: the telescope for astrophysical phenomena on large-distance scales and the microscope for biological phenomena on small-distance scales. As an object of observation, technology reveals phenomena that are intrinsic to the workings of the technology as it harnesses phenomena in its interaction with the world: on a large scale, airplanes reveal hydro-dynamic phenomena as they perform their intended function of flight, and, on a small scale, laser-atom coolers reveal light-matter interaction phenomena as they perform their intended function of atom cooling. Both as tool and object, as technologies grow in power, their ability to re-veal phenomena and patterns in those phenomena also grows in power, and their potential impact on the scientific method, of which the finding of facts is the first step, grows in power.

Two Types of Fact-Finding

As hinted at above, fact-finding is both precursor and follow-up to explanation-finding, and thus comes in two types. The first type, a follow-up to explanation-finding, is directed and intended to hypothesis-test—to test interim theory, to find facts that have been pre-dicted but not yet found. The second type, a precursor to explanation-finding, is open-ended, not intended to test any particular hypothesis but simply to observe phenomena and discover new patterns. Both types are important, and we discuss them in turn.

The first type of fact-finding, more directed and less open-ended, we call "hypothesis testing": the use of technology to elicit observa-tions about phenomena intended to test particular scientific theories or hypotheses. Hypothesis testing is exemplified by the Michelson-Morley experiment, which tested the hypothesis that the speed of light should depend on its direction relative to the motion of ether, a sub-stance then thought to fill vacuum. Many big science experiments are, in part, of this type: LIGO (the Laser Interferometer Gravitational-Wave Observatory), in part intended to measure the gravity waves predicted by Albert Einstein's theory of general relativity, and the LHC (Large Hadron Collider), in part intended to find the Higgs boson predicted by the standard model of particle physics. Directed fact-finding is also that most closely associated with the scientific method and, because it can have clear goals and milestones, the kind of fact-finding that most appeals to science-funding agencies. It is a powerful filter for eliminating nonsense—for eliminating theories that cannot be confirmed (or whose consistency with nature cannot be made more likely) through experiment.

But this first type of fact-finding, as important as it is, can take us only so far. The more open and less directed type of fact-finding is just as important. This second type, what we call "open-ended phenomena elicitation" is the use of technology to elicit observations about phenomena with little scientific preconception of what one might find (D. D. Price, 1984). This open-ended type is often the source of the truly new and unexpected, because it has not yet gone through the gauntlet and strong filters of prior fact-finding, explanation-finding into theory, and predictions of new facts consistent with that theory. Unexpected phenomena present themselves whether or not a particular phenomenon is purposefully being sought or a hypothesis is purposefully being tested. In the words of Herbert Simon, winner of the 1978 Economics Nobel Prize (Simon, 2001, p. 28):

> In one [type], we already have a general theory. We deduce some new consequences from the theory and then design an experiment to test whether these consequences are supported empirically. This is the pattern of what Kuhn called 'normal science.' In the other [type], there are some phenomena of interest—bacteria, say—but at the moment we have no hypothesis about them that we wish to test. Then we can design experiments, or merely opportunities for observation, that may reveal new and unexpected patterns to us. Having found such patterns, we can search for explanations of them.

The basic human desire, as old as the desire to climb to the top of a ridge to get a better view of the valley below, is simply to see more and farther.

Interestingly, although these two types of fact-finding are quite different, they are sometimes treated similarly. Open-ended phenomena elicitation is not scientific hypothesis-testing—there is no scientific theory that is being tested—even if it is often cast in that way. One might think, for example, that the conjecture "if we investigate a certain phenomenon with a higher-resolution microscope, we will see something new" is a scientific hypothesis. It is not. Instead, it might best be viewed as an engineering hypothesis—a form intended to fulfill the function of fact-finding through open-ended phenomena elicitation. That said, such engineering hypotheses are extremely important: discovering the new and surprising fact is made more likely the better the conjecture, born of deep craft and intuition, as to where one should look to find the new and surprising.

Open-Ended Phenomena Elicitation

Though open-ended phenomena elicitation is crucial to the finding of new and surprising scientific facts, it has a large element of serendipity. This is particularly so when the new phenomena or patterns in phenomena emerge completely unintentionally, as when a technology is being used in its normal use environment, and the alert scientist notices an unusual pattern in its behavior. When Arno Penzias and Robert Wilson noticed unusual noise in their microwave receivers, for example, their resulting investigations led to their discovery of the 2.7K cosmic microwave background (and to the awarding of half of the 1978 Physics Nobel Prize). Just the "fact" that an artifact or process fulfills a certain function, that it "works" in the way it does, can itself be an interesting scientific fact. This we might call an exaptation event: the technology is fulfilling some intended and human-desired *practical* function, but the user or observer is alert to an alternate but also human-desired *scientific* function.

But just because open-ended phenomena elicitation has a large element of serendipity does not mean it cannot have some degree of intentionality and purposefulness. If observations are made as a technology is intentionally used over a wider-than-normal range of performances or use environments, or observations are made of a technology modified so as intentionally to reveal new phenomena, then nature is more likely to provide observations difficult to explain with existing science. The ability to extrapolate phenomena *outside* of normal performance or use environments is much more uncertain than the ability to interpolate phenomena *inside* of normal performance or use environments, and thus more likely to reveal the new and surprising.

And herein lies the resolution to the common but mistaken belief that $\dot{S}$ must not be entangled with $\dot{T}$, lest it become $\dot{S}$ of a development rather than research nature, as we will elaborate in Chapter 3.

On the one hand, it can, surely, and some of what is called the "engineering sciences" is useful but less surprising $\dot{S}$, precisely because it is concerned with interpolation. Because observations would typically be made over a narrow range of performance or use environments, the observations will be of patterns in that narrow range and are likely to be explainable using existing science. Indeed, due to its closeness to technologies with narrow and specific performance needs, solid-state physics has been a particular example of the mistaken prejudice against entangling science with technology. Wolfgang Pauli, winner of the 1945 Nobel Prize in Physics, famously said, "Festkörperphysik ist eine Schmutzphysik" (solid-state physics is the physics of dirt), and Murray Gell-Mann, winner of the 1969 Nobel Prize in Physics, famously dubbed solid-state physics "squalid-state physics" (Natelson, 2018). Engineering a technology so that it improves performance and better fulfills some function under normal operating conditions may very well lead to a new observation, but likely only incrementally new.

On the other hand, engineering a technology so that it reveals new phenomena outside of normal operating conditions might very well lead to an unexpected observation. It does not take much to expand the range of observations a new technology can reveal, if one is so inclined, and to thus make it a rich source of raw material and patterns for scientific explanation. Wind tunnels enabled airplane shapes to be exposed to conditions far removed from normal use environments and thus led to new observations and new theory.

Indeed, instrumentation that enables seeing better and with higher resolution is a fruitful strategy of many great scientists. Galileo did not know what he would find but had good reason to believe he would find something new and interesting if he could improve his telescope to see farther. Antonie van Leeuwenhoek and Robert Hooke did not know what they would find but had good reason to believe they would find something new and interesting if they could improve their microscopes to see nearer. Examples like these abound, especially when the technology is relatively new and observing with the technology is like entering a new playground of phenomena just waiting to be found

(D. J. Price, 1986, pp. 237–253). X-ray diffraction, the Hubble telescope, high-resolution optical spectroscopy, electron microscopy—all of these technologies enabled us to see unexpected phenomena. In some cases, we see so clearly the possibility of finding something new and interesting that we are willing to invest significantly in the instrumentation required (Galison, 1997)—one result being "big science" (Weinberg, 1961), which should actually be called "big engineering in service of science" or "big engineering in service of fact-finding."

Taking advantage of the intentional use of technology for open-ended phenomena elicitation does, however, require a shift away from the goal of improving the technology for some practical human-desired function and toward the very different goal of eliciting new phenomena and new patterns of phenomena for a scientific purpose. This is a subtle but crucial shift in mindset and purpose. Louis Pasteur studied fermentation, apparently with an initial desire to improve this well-known technology for producing wine. Soon, however, he turned toward exploring fermentation in a wider range of operating environments, particularly those environments that led to normally unwanted products such as vinegar or lactic acid. Fermentation thus became a tool for studying chemical transformations more broadly rather than simply a target for process improvement. And, through use of this tool, he was able to show the existence of biological organisms capable of catalyzing anaerobic chemical transformation and to create an entirely new scientific paradigm for explaining such chemical transformation. Similar examples abound of scientists who intentionally went into areas where the phenomena studied were of "applied" interest then modified the object of study to become one of scientific interest. Pierre de Gennes's studies of liquid crystals, a form of matter of interest to flat-panel displays, led to his discovery of fundamental similarities between phase transitions in liquid crystals and superconductors (Plévert, 2011), a discovery that contributed to his winning the 1991 Nobel Prize in Physics.

Explanation-Finding

Having found facts, we are launched into the second mechanism of the scientific method: *explanation-finding*, the finding of explanations for facts—whether the explanations are qualitative or quantitative, whether they pertain to the physical or living worlds, whether they are in the process of debate hence highly provisional, whether they are widely accepted and part of conventional wisdom, or whether they are driven by long-known mysteries or by sudden surprises.

The process of explanation-finding has two pieces—the finding of a *possible* explanation and the verification of that possible explanation (the matching of explanation with fact).

Finding of Possible Explanation

We start with the finding of possible explanation: the creation of possible theories to explain facts or shallower theories. This is the taking of observed patterns in phenomena and constructing theories that might explain them—ideally, as discussed earlier, explanations as parsimonious and causal as possible. This is sometimes the brilliant stroke in science that collapses a mountain of data and patterns into a simple equation that, as the Nobel-prize-winning physicist Jean Baptiste Perrin says, can "explain the complex visible by some simple invisible" (Perrin, 1913). Such a brilliant stroke is exemplified by the famous equation from Albert Einstein's theory of special relativity ($E = mc^2$), which "made sense" of a wide range of then-contradictory facts and existing theories. It is also exemplified by Charles Darwin's theory of evolution, which "made sense" of and made coherent a wide range of then-disconnected facts ranging from observations of animal and plant breeding to observations of diverse natural species on remote islands.

We distinguish two important characteristics of the finding of possible explanations: first, that finding possible explanations is an act of imagination not of logic, so it cannot be predicted or planned; and, second, that the possible explanations we find and the theories we construct unavoidably influence, often strongly, fact-finding.

The first characteristic of the finding of possible explanations is that it is an act of imagination. Finding possible explanations is not an act of deductive reasoning whereby from axioms or first principles thought to be true one deduces what follows. It is not even an act of inductive reasoning, whereby from a limited set of observations one infers general patterns. Instead, it is an act of *abductive* reasoning, in which the patterns are given, and one guesses at possible causal explanations for them.

Despite what many believe, theory cannot be deduced directly from phenomena. As articulated eloquently by Albert Einstein with respect to the great explanation finders, including Isaac Newton, of the seventeenth century (Einstein, 1934, p. 166):

> Newton, the first creator of a comprehensive and workable system of theoretical physics, still believed that the basic concepts and laws of his system could be derived from experience; his phrase "hypotheses non fingo" [I frame no hypotheses] can only be interpreted in this sense. . . . [In fact,] . . . the scientists of those times were for the most part convinced that the basic concepts and laws of physics were not, in a logical sense, free inventions of the human mind, but rather that they were derivable by abstraction, i.e., by a logical process, from experiments. It was the general Theory of Relativity which showed in a convincing manner the incorrectness of this view. For this theory revealed that it was possible for us, using basic principles very far removed from those of Newton, to do justice to the entire range of the data of experience in a manner even more complete and satisfactory than was possible with Newton's principles. But quite apart from the question of comparative merits, the fictitious character of the principles is made quite obvious by the fact that it is possible to exhibit two essentially different bases, each of which in its consequences leads to a large measure of agreement with experience. This indicates that any attempt logically to derive the basic concepts and laws of mechanics from the ultimate data of experience is doomed to failure.

The inability to deduce theory from phenomena is not to downplay the importance of the observed patterns of phenomena themselves; these are, of course, of the utmost importance. But these observed patterns, derived from experience, are only a test for an explanation; one must come up with that explanation in the first place; and that "coming up with an explanation," that generative act in all science, is to make guesses—acts of pure imagination that are neither predictable nor plannable. Likewise, the importance of imagination is also not to downplay the importance of deep domain-specific knowledge alongside imagination. Theory and imagination stumble and make wrong turns, often. If one includes all the conjectures theorists have made that never became public, one might conclude that a negligible fraction of conjectures survives and that the imagination is almost always wrong—particularly when it is attempting to explain observations that are striking and seemingly inexplicable. Deep domain-specific knowledge is necessary to weed out instances where imagination has gone wrong.

The second characteristic of the finding of possible explanations is that these possible explanations themselves influence fact-finding. Patterns and explanations go hand in hand; sometimes patterns lead to explanations, but sometimes explanations lead to noticing patterns that one didn't notice before. "I'll see it when I believe it" is just as important as "I'll believe it when I see it."

The influence of explanation-finding on fact-finding has a powerful upside: how many times in the laboratory does one observe something that makes no sense, and on a deeper look it turns out an error was indeed made in setting up the observation to begin with. The Millikan oil drop experiment is a classic example for which data was discarded (oil drops that, from their velocities in combined gravitational and electric fields, did not appear to have integer multiples of electron charge), likely because of Robert Millikan's preexisting belief that electrons and hence electronic charge were indivisible (Niaz, 2000). Indeed, use of explanation-finding in filtering observation is very important when data is complex and fact difficult to discern. So-called grounded theory can only take one so far in making sense of complex data—data must be simplified via biases and preexisting explanations, even as one must be willing to challenge those biases and preexisting explanations.

The influence of explanation-finding on stylized-fact-finding can also have a powerful downside: "one cannot escape one's mental map, the map that tells the researcher what is interesting, salient, or significant, and what is not" (A. C. Lin, 2002, p. 193), even if one's mental map is outdated or incorrect. Scientific progress, Thomas Kuhn argued, is not simply the continual discovery of new facts duly explained. Instead, progress changes what counts as "facts" in the first place. When reigning theories are replaced by incommensurable challengers, the purported facts are redescribed according to new and incompatible theoretical principles. Thus, purported facts are not simply a recital of observations; rather, they are the interpretation of observations on the basis of all the theories the scientist regards as true. Millikan's observation of oil droplets with noninteger-multiples of electron charge is attributed to experimental noise rather than to "subelectrons" (Niaz, 2000).

Ultimately, a balance between facts and possible explanations must be struck. In the social sciences, to avoid the subjectivity of human beings, the spillover of the finding of possible explanations into fact-finding is often mistrusted. In the physical sciences, the spillover of the finding of possible explanations into fact-finding *is* often trusted: the stunning experimental result of cold fusion turned out not to be correct; existing theory was, instead. Incompatibility of experiment with theory does not necessarily mean the theory is wrong—there may be a subsidiary interpretation that is wrong, and the overall theory might still be fixable and right. Confirmation bias, persistence in the face of implausibility, can be important. Albert Einstein famously said that if the experiment hadn't supported his theory, he would have still believed his theory and distrusted the experiment. Nonetheless, one has to be self-critical. One must be as wary of overconfidence in proposed explanation as of overconfidence in proposed facts.

Verification of Explanation

We turn now to verification, the working-out of the consequences of explanations so as to verify that the observed facts are, in fact, consequences of the explanation. Why is this important? Because even though explanations are sometimes simple, they are often only deceptively so. For example, we observe a ball being thrown to first base, and we're told that the observed motion of the baseball is due to the acceleration of that object being equal to the force exerted on that object (by the pitcher) divided by its mass ($a = F/m$). But the consequences of even such simple explanations can be exceedingly difficult to work out: calculating the motion of the baseball required the invention of calculus, an entirely new field of mathematics. The consequences may even be intractable to analytic mathematics, with numerical simulations required instead. The three-body problem in mechanics, in which each body exerts a force on the other two, has famously resisted analytic mathematical solution despite the ease with which it can be stated.

Just like fact-finding, verification-finding depends heavily on technology. But, whereas fact-finding depends heavily on physical technology to observe physical observations that might become fact, verification-finding depends heavily on calculational technology for deducing the consequences of explanation. Calculational technology includes analytical mathematics, and, for centuries, this was the main available option. It also includes, especially with the advent of powerful computers, algorithmic and numerical techniques.

Also just like fact-finding, verification-finding comes in two types: verification seeking and verification discovery. By "verification seeking," we mean the seeking (and development) of calculational techniques for the purpose of working out the consequences of specific explanations: the invention of calculus for the purpose of working out the consequences of acceleration on the motion of bodies, or the invention of density-functional theory for the purpose of calculating the electronic structure of many body systems. By "verification discovery," we mean the open-ended exploration of calculational techniques without a particular explanation in mind to verify but that might someday enable verification of explanation. Many domains of mathematics, initially thought to be "pure," have come to have an "applied" component as they were gradually used to work out the consequences of scientific explanations: group theory applied to virtually all physical phenomena in which symmetry plays a role, or linear algebra applied to quantum mechanics. Many computational algorithms, initially developed to fulfill some practical human-desired functions, were later "discovered" to be useful for scientific verification: the fast Fourier transform, developed to analyze data from sensor arrays, is now used widely to simulate scientific explanations involving time- and frequency-dependent phenomena.

Note that there is a blindness and uncertainty that accompanies verification of explanation. After an "aha" moment in which a possible explanation comes into view, the calculation of the consequences of that explanation can be difficult and time-consuming—and during that time there is often a sense of excitement and trepidation over whether the consequences will match observed fact. A similar blindness and uncertainty accompany verification of the engineered forms discussed earlier. After an "aha" moment in which a design for an engineered form comes into view, the physical instantiation of that engineered form can be difficult and time-consuming—and during that time there is often a similar sense of excitement and trepidation over whether the consequences will match desired function.

The need in both cases for verification, born of blindness and uncertainty, is an indication of how both represent hypotheses: a proposed scientific explanation is a *scientific hypothesis*, before its matching with fact has been verified; while a technological form or a design for such a technological form is an *engineering hypothesis*, before its matching with desired function has been verified. If the verifications are calculational, the tools for verification can be very similar, involving deep mathematical and algorithmic computation—with scientific verification sometimes more elegant because simplifications are sought and engineering verification sometimes messier because the real world and its complications must be taken into account. And if the engineered form embodies the phenomenon that a scientific hypothesis proposed to predict, then the scientific and engineering hypotheses can be considered one and the same. As will be discussed later in this chapter, each of the successive engineered forms of the transistor embodied the phenomenon that the current scientific hypothesis predicted would be observed, so matching or not matching of those forms with desired transistor function simultaneously confirmed or disconfirmed that prevailing scientific hypothesis along with the engineering hypothesis that functional transistor gain might be found. Human motives are often mixed and contain both a scientific *and* engineering character—all the more reason, as discussed in Chapter 4, not to isolate science and technology or source of inspiration (curiosity or practical application) from each other in research.

Generalizing

Having found a library of explanations, we are now launched into the third mechanism of the scientific method: *generalizing*, the extension of theory, originally intended to explain one set of facts, to other facts it was not intended to explain. Einstein's theory of special relativity, for example, was intended to explain the constancy of the speed of light, but ended up also explaining energy release during nuclear fission/fusion. In a sense, any new theory must pass two tests. The first test, what might be called "particularization," is an internal one—the theory must first explain the particular phenomena it is *intended* to explain. The second test, what might be called "generalization," is an external one—the theory must also explain something *else,* something more general than the phenomena it was intended to explain. As articulated by Richard Feynman (Feynman, 1974):

> When you have put a lot of ideas together to make an elaborate theory, you want to make sure, when explaining what it fits, that those things it fits are not just the things that gave you the idea for the theory; but that the finished theory makes something else come out right, in addition.

Why is this second test, that the theory generalizes, so important to the scientific method? It is important for two reasons.

The first reason generalization is important is that the more generalizable the theory, the more broadly applicable it is and the more fertile it will be with respect to what Thomas Kuhn called "articulation of theory" (Kuhn, 1962). The further a theory can be articulated, the more scientists will be interested in adopting the theory. Ultimately, scientific principles are more powerful the more they generalize across previously seemingly orthogonal phenomena: Antoine Lavoisier's conservation of mass applying to all chemical reactions; James Clerk Maxwell's equations applying both to radio waves and light; and the BCS (Bardeen-Cooper-Schrieffer) theory, which not only explained superconductivity but also generalized to new phenomena such as magnetic flux quantization and the Josephson effect. The more generalizable a theory, the more embedded it becomes in the "seamless web of knowledge" that we will discuss in Chapter 2, and the more broadly applicable and powerful.

Interestingly, humans are adept at generalizing but artificial intelligence (AI) is not—yet. We humans tend to extrapolate explanations beyond their original bounds, to the point of *overgeneralizing*—a common human cognitive bias. But artificial intelligence is not yet able to do even elementary generalization (Marcus, 2018). Such "transportability" of explanations would be an important step from simple to general AI. In simple AI and classic machine learning, a cognitive entity "observes" a training set (for example, images categorized as dogs or cats). The entity makes sense of that training set by optimizing the numerical weights in its neural network. Any new image that is similar to an image in its training set, it can identify accurately, because it is "interpolating" its model within the narrow knowledge space from which its observations come. In more general AI, a cognitive entity might make the ansatz, or guess, that its model (or some piece of its model) is generalizable outside the narrow knowledge space from which its training observations came. It might propose to make an observation that is an extrapolation outside of that narrow knowledge space. It might even construct an artificial image that is purposefully well outside of the narrow knowledge space, classify it using its generalized model, and ask the world whether it has classified correctly or not. The cognitive entity is generalizing, and the further the extrapolation is from the narrow knowledge space, the more striking the generalization and the bigger its leap in knowledge. Indeed, this is one reason why causality is emerging as an essential ingredient for general AI. It is when "hard" causality, which is more generalizable, is found—rather than "soft" correlation, which is not as generalizable, is found—that explanations might generalize across environments (Marcus, 2018; Pearl & Mackenzie, 2018).

The second reason generalization is important is that, the more generalizable a theory, the more likely it is to be true because it is more easily falsifiable. Echoing Feynman, one important outcome of generalization is the finding of hypotheses, the prediction of some not-yet-observed phenomenon beyond the phenomena for which the theory was originally constructed. Moreover, the more implausible the predicted phenomenon and/or the more disparate or "orthogonal" the predicted phenomenon from the original phenomenon, the more likely the predicted phenomenon will not be found—hence, if found, the more likely that the theory is true. An example of such an implausible prediction was that of Simeon Denis Poisson in 1818. Based on Augustin-Jean Fresnel's new wave theory of light, Poisson famously predicted an on-axis bright spot in the shadow of a circular obstacle. When Dominique-François-Jean Arago did the experiment, he found, against his own intuition, the predicted "spot of Arago," thereby making the new wave nature of light much more credible (Harvey & Forgham, 1984). An example of the prediction of multiple phenomena orthogonal to each other are flashes of light and an accompanying but delayed clap of thunder. When one sees the one and hears the other, it is likely that there was lightning because visual and auditory senses are so disparate, their triggering in particular ways so precisely delayed in time is unlikely to just be coincidence. Indeed, if a theory or explanation is based on a very rich problem set (a set of very disparate observations), so that the internal test is already extremely severe, then the likelihood of truth is high regardless of any subsequent external test. Like Einstein, Maxwell said his theory *had* to be right because it was consistent with too many disparate observations unexplainable in any other way—the internal test was extremely severe. Nonetheless, subsequent external tests, born of generalization, ultimately serve as ongoing arbiters of truth.

1.3 Cycles of $\dot{S}$ and $\dot{T}$: The Transistor and the Maser / Laser

We have seen that the creation of new science depends on existing science and technology, as does the creation of new technology. $\dot{S}$ and $\dot{T}$ stand on the shoulders of existing S and T, and science and technology are intricately woven into each other in powerful (and beautiful) technoscience feedback cycles.

Indeed, referring to new technology as invention and new science as discovery, their mutual ratcheting forward can be thought of as occurring in mutually reinforcing cycles of invention and discovery (Narayanamurti & Odumosu, 2016). Technology fulfills a key function for science in that it enables experimentation. Experimental technologies establish the facts that initiate the scientific method, while calculational technologies establish the consequences of explanation so that explanation can be verified against fact. Most importantly, as experimental and calculational technologies get more sophisticated (for example, high-resolution electron microscopes or highly parallelized supercomputers), facts at higher levels of precision can be established, and explanations can be verified or "unverified" with greater confidence. As science gets more sophisticated, the modeling of the phenomena on which technology is based becomes more sophisticated. Functions that are impossible to fulfill because they contradict known explanatory principles can be eliminated; forms can be designed with greater confidence before they become instantiated in the real world.

Moreover, the degree of internal amplification as the technoscience cycles are traversed increases with the degree to which each mechanism is balanced toward exploration and surprise. New and surprising forms (for example, functional magnetic resonance imaging) widen the range of phenomena that can be observed, giving more opportunity for new and surprising facts (brain activity response to various stimuli) to be found. Explaining new and surprising facts increases the opportunity for improving forms that harness phenomena associated with those facts. In other words, new and surprising scientific knowledge amplifies the creation of new and surprising technology, just as new and surprising technological knowledge amplifies the creation of new and surprising science.

The following case studies of two of the twentieth century's iconic technoscientific advances, the transistor and the maser / laser, detail the ways in which they benefited from the powerful feedback cycles of science (discovery) and technology (invention).

The Transistor

We begin with the transistor. Its antecedents, both scientific *and* technological, were in the air during the years leading to its invention. On the science side, the early twentieth century saw the emergence of quantum mechanics, including our understanding that electrons (and, indeed, all matter) can be thought of as both particles and waves, and of the theory of such electrons, treated as waves, in solids. The theory of electrons in a certain kind of solid, semiconductors, was also emerging—a theory far removed from "common sense" and understandable only through new and increasingly sophisticated solid-state physics. On the technology side, the early twentieth century saw the emergence of vacuum tubes, in which electrons are emitted from a cathode into a vacuum, then collected by an anode. By the 1940s, vacuum tubes had created a new discipline of electronics, with applications including radio, television, radar, sound recording and reproduction, and long-distance telephone networks.

There was thus a sense of significant opportunity for exploring semiconductor science and technology, as well as recognition that exploring semiconductor science and technology *together* would prove especially fruitful. Toward the end of World War II, Mervin Kelly, then executive vice president of Bell Labs, created a solid-state physics group whose mission he articulated thusly (Riordan & Hoddeson, 1998, p. 116):

The quantum physics approach to [the] structure of matter has brought about greatly increased understanding of solid-state phenomena. The modern conception of the constitution of solids that has resulted indicates that there are great possibilities of producing new and useful properties by finding physical and chemical methods of controlling the arrangement and behavior of the atoms and electrons which compose solids.

Employing the new theoretical methods of solid-state quantum physics and the corresponding advances in experimental techniques, a unified approach to all of our solid state problems offers great promise. Hence, all of the research activity in the area of solids is now being consolidated in order to achieve the unified approach to the theoretical and experimental work of the solid-state area.

Co-led by William Shockley and Stanley Morgan, the new group was composed of a mix of those with scientific and engineering bents—reflecting Shockley's interest in establishing a culture in which both were valued. In no small part due to the inspiration of Mervin Kelly, Shockley was intensely interested in creating a solid-state amplifier, the semiconductor equivalent of the vacuum triode, while *at the same time* he was interested in developing a comprehensive theory of semiconductors. As he would later elaborate in his Nobel lecture (Shockley, 1956, p. 345):

> Before leaving the subject of research in industry, I would like to express some viewpoints about words often used to classify types of research in physics; for example, pure, applied, unrestricted, fundamental, basic, academic, industrial, practical, etc. It seems to me that all too frequently some of these words are used in a derogatory sense, on the one hand to belittle the practical objectives of producing something useful and, on the other hand, to brush off the possible long-range value of explorations into new areas where a useful outcome cannot be foreseen. Frequently, I have been asked if an experiment I have planned is pure or applied research; to me it is more important to know if the experiment will yield new and probably enduring knowledge about nature. If it is likely to yield such knowledge, it is, in my opinion, good fundamental research; and this is much more

important than whether the motivation is purely esthetic satisfaction on the part of the experimenter on the one hand or the improvement of the stability of a high-power transistor on the other. It will take both types to "confer the greatest benefit on mankind" sought for in Nobel's will.

During the five-year period (1944–1949) following the creation of this group an intense back-and-forth between scientific and technological achievements took place that culminated in the simultaneous invention of the transistor and discovery of the transistor effect. In the following pages, we discuss the six most important of these achievements, the ones highlighted in gray in Figure 1-3.

1946: The Null Thin-Film FET (Field-Effect Transistor)

The first achievement began with what might be called the "null thin-film FET." It was motivated by Shockley's form-finding insight, based on the semiconductor science of the time, that a field-effect transistor was a realistic route to an amplifying device that could replace the bulky and unreliable triode tube. The idea was that the charge and conductance of a thin semiconductor film acting as one of a pair of capacitor

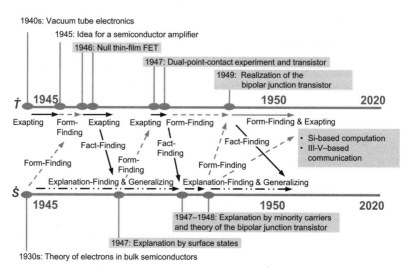

FIGURE 1-3. Cycles of $\dot{S}$ and $\dot{T}$ as exemplified by the transistor, including scientific and technological knowledge advances.

plates, could be modulated by charging or discharging the capacitor—
essentially by applying a perpendicular electric field to the thin film.
That form-finding insight, combined with the deep craft associated
with evaporated and other thin-film layers, was instantiated by Shockley and a number of collaborators in a variety of "forms" intended to
demonstrate the "functional" effect experimentally. None succeeded:
sometimes the sign of the effect was reversed, and even when not reversed its magnitude was orders of magnitude smaller than theoretically
expected. Nonetheless, the null result was not merely a failure to produce an amplifier, but also it was exapted into a significant new fact, a
null thin-film FET, ready for scientific explanation.

1947: Explanation by Surface States

Progress toward understanding the null thin-film FET was at a standstill
until John Bardeen found an explanation based on surface states that
could both localize and immobilize induced electronic charge at the surface, thus rendering them unavailable for enhanced conduction, as well
as screen interior electronic charge and render them as well unavailable
for enhanced conduction. Bardeen's concept of surface states immediately explained the newly observed phenomenon. Moreover, his concept
was both a generalization and generalizable: it was the surface analog of
other theoretical concepts (such as donors and acceptors, which were
known to apply to the interior of a semiconductor), and it explained
other previously mysterious phenomena, including the rectifying characteristics of semiconductors when contacted either by metal points,
known as cat's whiskers, or by other semiconductors (Bardeen, 1957).
The generalizability of Bardeen's concept immediately gave it great credibility and led to new ideas on how one might "engineer away" the deleterious effects of surface states.

1947: The Dual-Point-Contact Experiment and Transistor

What followed was a flurry of engineered new "forms": artifacts intended
both to test the new theory of surface states and to neutralize those surface states so that a field effect transistor could be built. A key breakthrough came from another field: the suggestion, on November 17,
1947, by Robert B. Gibney (a physical chemist who had been recruited to

round out the scientific expertise of the semiconductor group) to use a standard physical chemistry tool, an electrolytic solution. Such an electrolyte could be used to apply a strong electric field in the immediate vicinity of (while being electrically insulated from) a point contact on the semiconductor. Gibney and his team quickly found that the strong electric field did indeed alter the density of surface charges, and some amplification was found, though only at very low frequencies due to what they realized were the long time constants of the electrolytes. Experiments were devised to duplicate the effect, but with dual metal-point-contacts extremely closely spaced so that the proximity between contacts would nearly equal the proximity between contact and electrolyte. After a number of attempts, one worked: contacts were made by evaporating gold on a wedge, then separating the gold at the point of the wedge with a razor blade to make two closely spaced contacts on n-type germanium. Amplification was observed, both voltage and power, and the point-contact transistor was born.

Though there was clear intentionality associated with the engineered form of the dual-point-contact device, the observed amplification was entirely serendipitous. It was a strong enough effect to be brought to the attention of Bell Labs management in the famous "day before Christmas Eve" demonstration of December 23, 1947; but it did not follow the expectation of field-effect theory. The result was dissimilar to the null thin-film FET result in that it was a positive rather than negative result for an amplifier, but it was similar in that it could be exapted into a puzzling new scientific fact needing a new explanation.

1947–1948: Explanation by Minority Carriers and the Theory of the Bipolar Junction Transistor

The puzzling new fact of amplification by the dual-point-contact transistor was made sense of quickly: the amplification effect was not due to *majority* carriers (as first thought), but was instead due to *minority* carriers. In this case, holes in the n-type germanium were injected from one contact into the semiconductor, then collected by the other contact out of the semiconductor. Although minority carriers were known to exist at least since the earlier work of Wilson in the 1930s, their important role in an amplification device was an unexpected generalization of that earlier work. Once the generalization was accepted, though, it paved the way to a detailed working out, by William Shockley, of a comprehensive theory of electrons and holes in semiconductors (Shockley, 1950).

1949: Realization of the Bipolar Junction Transistor

Out of this theory came yet another form-finding insight: that the minority-carrier amplification phenomenon could be harnessed much more efficiently in a much more manufacturable device geometry. Thus, the bipolar junction transistor was born: a sequence of n-p-n layers composed of two p-n junctions placed back-to-back. The shared p "base" layer of those two junctions is thin, so that almost all the minority carriers "emitted" into the p layer by one forward-biased p-n junction are "collected" from that p layer by the other reverse-biased p-n junction. The tiny fraction of carriers that aren't collected is an extremely small emitter-base current that controls (and is "amplified" into) the much larger base-collector current. Minority carriers are crucial: if only majority carriers could conduct in the p-type base, then the transistor would not work. But it took the semiaccidental discovery of the point-contact transistor to focus attention on minority carriers and trigger the form-finding insight of the much more efficient and manufacturable bipolar junction transistor.

1950–2020: Longer Time Horizon Follow-On Semiconductor Achievements

The "tight" S and T interactions between 1944 and 1949 leading to the transistor and the transistor effect exercised every one of the mechanisms of the technoscientific method discussed earlier—with fact-finding and form-finding serving as the principle "cross" mechanisms by which T influenced $\dot{S}$ and S influenced $\dot{T}$. The transformational nature of the work was quickly recognized, and Bardeen, Brattain, and Shockley after only seven years were awarded the Nobel Prize in Physics. The transistor was only the starting point, of course. After a subsequent seventy years of continued S and T interactions, with no sign of abating, we now know that the transistor spawned an entirely new field of semiconductor electronics, which has been pivotal for many technologies, including computation and communication.

The Maser and Laser

We turn now to the laser ("light amplification by stimulated emission of radiation") and its predecessor, the maser ("microwave amplification by stimulated emission of radiation"). Just as for the transistor, the antecedents of the maser were in the air during the years leading to its invention. Again, importantly, these antecedents were both scientific *and* technological. On the science side, the early twentieth century had seen emerge an understanding of the electronic structure of atoms and molecules, of the energy (rotational, vibrational, and electronic) states associated with those structures, and of the transitions between those energy states. Albert Einstein had identified the various ways in which photons could mediate transitions between these energy states, including in one unanticipated new way: stimulated emission (Einstein, 1917). On the technology side, the early twentieth century had also seen the emergence of radar (radio detection and ranging), in which pulsed electromagnetic waves in the radio and microwave (MHz-GHz) frequency range were reflected off objects to accurately determine their distance. Radar technology development accelerated during World War II, as did development of all the components necessary to create and manipulate radio and microwave radiation: transmitters for generating the radiation (klystrons and magnetrons), waveguides for guiding the radiation, antennas for emitting the radiation into (and collecting the radiation from) free space, and filters and receivers for selecting and detecting the radiation.

Unlike for the transistor, for which researchers and senior management at Bell Labs and elsewhere recognized the potential for new and useful semiconductor science and technology, little such recognition existed for the maser, very likely because there was no device analogous to the maser, while there *was* a device analogous to the transistor—the vacuum triode, albeit operating on different principles than the transistor. Thus, the quest for the maser in the late 1940s and early 1950s was almost entirely due to the vision of a handful of researchers: Charles Townes (then at Columbia), Nikolay Basov and Alexander Prokhorov (then at the Lebedev Physical Institute), and Joseph Weber (then at the University of Maryland). Even these researchers were discouraged at the time by others: by Townes's management at Bell Labs (which caused him to relocate to Columbia) (Townes, 1999, pp. 43–46) and even by

Prokhorov's students at the Lebedev Physics Institute, whose desire to do research on other topics caused him to destroy with a hammer the instruments they needed for research on those other topics (Graham, 2013).

The motivation for the maser, unlike for the transistor, was not for any specific practical application. It was to advance science—to explore ways in which resonances between energy levels in molecules could be useful for improved microwave technology and then in turn useful for ever more accurate microwave spectroscopy. Indeed, soon after Theodore Maiman (at Hughes Research Laboratories) had demonstrated the first laser, in 1960, his assistant joked that the laser was "a solution looking for a problem" (Hecht, 2010), and in Townes's Nobel lecture of 1964 (Townes, 1965), he did not even mention what ended up being the most important use of lasers: communications. Nonetheless, there was a sense that the new functionality of coherent radiation at higher frequencies and shorter wavelengths than ever before, produced in a completely new manner using resonant molecules as microwave circuit elements, would lead to the new and unexpected.

It was during the twelve-year period between 1948 and 1960 that back-and-forth scientific and technological achievements took place that culminated in the inventions of the maser and laser, the most important of which are highlighted in gray in Figure 1-4.

1948: Idea of Population Inversion

While Einstein had early on identified stimulated emission as a new way in which photons could mediate transitions between energy states, it was long thought that stimulated emission would always be dominated by stimulated absorption ("absorption," for short). In any set of energy levels in thermal equilibrium, the higher energy levels would be less populated than the lower energy levels, so stimulated emission, proportional to the population in the higher levels, would be dominated by absorption, proportional to the population in the lower levels. Amplification would then be impossible. Thus, the outstanding *functional* problem for generating amplification, in the late 1940s, was how to achieve a higher population in the higher energy than in the lower energy levels—in other words, how to achieve population *inversion*.

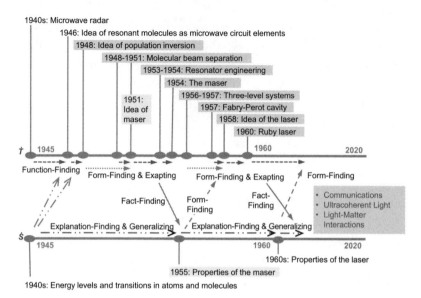

FIGURE 1-4. Cycles of $\acute{S}$ and $\check{T}$ as exemplified by the maser and laser, including scientific and technological knowledge advances.

1948–1951: Molecular Beam Separation and the Idea of the Maser

The seeming impossibility of significant population inversion probably prevented it from being pursued very intensely. A fortuitous juxtaposition of knowledge and experience, however, enabled Charles Townes, in 1951, to exapt a number of technological "forms" to serve the new functionality he sought in a maser. First, Townes was extremely familiar with ammonia (NH_3) and the charge distributions of its various energy states from his molecular spectroscopy studies. Second, he was familiar with microwave resonators and aware of how much population inversion would be needed to support not just amplification but oscillation. Third, and of decisive practical importance, Townes was familiar with the deflection of beams of molecules with dipole and quadrupole moments in inhomogeneous magnetic and electric fields. He had just heard from a visiting German physicist, Wolfgang Paul, that the technique had recently improved to the point that the deflection might be used to separate (and population-invert) excited-state from ground-state molecules. In other words, molecular beam separation appeared to be an exaptable technique for creating a population inversion capable of amplification.

It was the integration of these three technological forms (ammonia energy states, microwave resonators, and deflection of molecules in magnetic and electric fields), that enabled Townes to engineer the "idea" of a practical maser, which he later described thus (Townes, 1999, p. 57):

> Answers were actually well known; they had been in front of me and the physics community for decades. Rabi, right at Columbia, had been working with molecular and atomic beams (streams of gases) that he manipulated by deflecting atoms in excited states from those of lower energies. The result could be a beam enriched in excited atoms. At Harvard, Ed Purcell and Norman Ramsey had proposed a conceptual name to describe systems with such inverted populations; they had coined the term "negative temperature," to contrast with the positive temperatures, because these "negative" temperatures inverted the relative excess of lower-level over upper-level states in equilibrated systems.

1952–1954: Resonator Form-Finding

One piece still missing from the design of a microwave oscillator, however, was the resonant cavity in which ammonia molecules would interact with the microwave field. The cavity needed to be highly reflective and low loss so that microwave radiation would build up via stimulated emission rather than be absorbed as it circulated inside the cavity—this seemed to imply enclosing the cavity as much as possible. At the same time, excited-state molecules needed to enter and exit the cavity through entrance and exit holes—but these holes were potential leakage paths for the circulating radiation. As Townes describes, the solution came in the following way (Townes, 1999, p. 66):

> One day, Jim Gordon opened the ends almost completely; that was what put it over the top. Without a ring in each end, we could get plenty of molecules through the cavity. Our worry over too much radiation leaking from the ends was unnecessary. Apparently, without the rings, the pattern of radiation in the cavity became simpler and more efficiently confined. The cavity was quite long, and radiation largely bounced back and forth

between the sidewalls, so not much leaked out the perfectly circular, large holes in the ends. Probably, the slitted rings, which had previously been fitted on the cavity, had neither been perfectly circular nor well enough connected to the cavity, thereby distorting the fundamental resonance pattern of radiation in the cavity and actually enhancing the loss of energy.

The solution was surprising and involved a healthy dose of serendipity and intuitive tinkering.

1954: The Maser and Its Properties

The successful experiments took place in Townes's laboratory from 1952–1954, drawing on deep science (molecular spectroscopy and stimulated emission), deep technology and craft (microwave electronics and instrumentation), and, as just mentioned, luck. It was simple: an evacuated rectangular metal box into which a tube introduced heated ammonia gas as a beam of molecules; inside the box a molecular focuser made of four parallel tubes arrayed in a square focused only the ammonia in excited energy states through a hole into the resonant cavity; after the molecules flowed beyond the resonant cavity, they were removed by either a vacuum pump or by a surface at liquid nitrogen temperature. The cavity was resonant at the ~24 GHz frequency associated with the strongest of ammonia's 1.25-centimeter wavelength transitions, a wavelength for which Townes knew a lot of useful manipulation techniques. In 1954, oscillations were observed, and the first maser was born.

Once the first maser was born, the microwave radiation it emitted could be measured and its properties studied. There was considerable uncertainty in particular over the frequency purity to expect from the radiation. The idea that the frequency would be extremely pure was dismissed at first blush by some of the leading physicists of the time, including Niels Bohr (Townes, 1999, pp. 69–72). But by building two masers and mixing their radiation, Townes found that the "beat" frequency was exceedingly pure; by implication, the frequencies of the radiation from both of the masers had to be exceedingly pure. The new "fact" of extreme frequency purity was then explained by the now-well-established Schawlow-Townes equation for the fundamental quantum limit to the line widths of both maser and laser radiation.

Shortly after researchers demonstrated the maser, new methods for generating population inversions were imagined—particularly since the molecular beam method then in use was cumbersome and not amenable to condensed-phase gain media. A breakthrough came in 1954–1956 from Nicolai Basov and Alexander Prokhorov (Basov & Prokhorov, 1955) and from Nicolaas Bloembergen (Bloembergen, 1956), an expert in transitions between electron and nuclear spin states in condensed phases. They envisioned three-level masers, in which microwave energy first excites electrons from their lowest (first) level to their highest (third) level, bypassing the intermediate (second) level. Depending on the rates of deexcitation via spontaneous emission, a population inversion can be obtained between the third and second, or second and first levels. Soon after, in 1957, Henry Scovil and George Feher at Bell Labs demonstrated a working maser in gadolinium ethyl sulfate based on this principle (Scovil et al., 1957). Three-level systems, a clever piece of form-finding based on deep scientific insights from spin resonance studies, thus enabled a new and much broader class of gain media than those compatible with molecular beam separation. Further, though it was not understood at the time, their concept would be easily exapted from masers to lasers, enabling gain media that were also compatible with lasers.

1957: Fabry-Perot Cavity

Because masers were basically a way to harness molecular resonances in service of extending microwave technology to higher frequencies and shorter wavelengths, it was natural for Townes to be thinking about extending even further, into the optical frequency domain. In other words, to extend the concept to optical masers, or what we now call lasers. Thanks to the three-level system breakthrough just mentioned, there were plenty of gain media to consider. However, form-finding the resonant cavity was a major concern. The leap from wavelengths in the mm-to-cm range to wavelengths in the tenths-of-μm range meant that the cavity would be thousands of times larger than the wavelength of the radiation, containing many resonant modes and not selective for any particular one. The answer came from Art Schawlow (then at Bell Labs), who proposed exapting a simple Fabry-Perot cavity: two plane-parallel mirrors with no sides. Without sides, there would be no off-axis oscillating modes—any off-axis photon would strike the end mirror at an angle and eventually "walk" itself out of the cavity rather than build up energy. Only those few modes that were exactly on-axis would survive and oscillate.

1958: Idea of the Laser

It was the integration (and exaptation) of these three technological forms (the maser, three-level systems, and the Fabry-Perot cavity) that enabled Schawlow and Townes to engineer the idea of a practical laser. In 1958, they published a paper that discussed, based on established scientific theories and likely engineering configurations, the feasibility and potential performance of a laser, or what they called then an "optical maser" (Schawlow & Townes, 1958). The paper was a landmark, presaging much of what was to be seen experimentally in subsequent years, and it epitomizes the use of scientific principles and insights to engineer a complex device.

1960: Ruby Laser

With the publication of Schawlow and Townes's paper, both scientific and engineering communities recognized its potential importance. Many groups entered the race to engineer a working laser, each with different gain media and different methods for creating a population inversion. The winner, in 1960, was Ted Maiman (Hughes Research Labs), whose gain medium was ruby and whose method for creating a population inversion was to slip a small ruby rod inside the coil of a flashlamp borrowed from the photographer's tool kit, all enclosed in a reflective cylinder. There was some skepticism that Maiman would succeed: Schawlow himself had decided ruby would not work, in part because some measurements had shown its red fluorescence was inefficient. But Maiman had made his own measurements and knew that ruby fluorescence was actually quite efficient; he also knew the material well from having used it to design a compact microwave maser. Maiman pressed ahead and succeeded.

The tight science and technology interactions during the 1948–1960 period that led to the maser and laser exercised all of the $\dot{S}$ and $\dot{T}$ mechanisms of the technoscientific method, with form-finding being the principle mechanism by which S influences $\dot{T}$ and fact-finding being the principle mechanism by which T influences $\dot{S}$. The ruby laser achievement opened the floodgates to lasers with a wide variety of gain media. And, though at first the applications were all scientific (using lasers as instruments to probe optical phenomena in materials as well as studying the physics and properties of lasers themselves), before long it became clear that lasers had revolutionary implications on, and could be exapted to, nonscientific applications as well. After a subsequent sixty years of continued science and technology interactions, with no sign of abating, we now know that the laser played and continues to play pivotal roles in the fields of optics and optoelectronics, including the fiber-optic communications technology that is the backbone of the modern Internet, not to mention opening up entirely new fields such as quantum optics, nonlinear optics, and ultrafast optical phenomena.

1.4 Recapitulation

To recapitulate, this chapter outlined our first stylized fact associated with the nature of research: that science and technology are two equally important but qualitatively different repositories of human knowledge and that these repositories coevolve deeply interactively to the benefit of each. Indeed, the degree of interaction is so intimate that it is tempting to blur the line completely and treat the two as a single entity called technoscience. Distinguishing between the two is useful, however, because it enables a finer view of the mechanisms by which they coevolve and a finer view of how these mechanisms can, as we will discuss later in Chapter 4, best be nurtured.

We also introduced a new nomenclature to define these mechanisms, along with a new unified framework for organizing them: the *technoscientific* method, composed of a "scientific" method and an analogous "engineering" method. The scientific method is comprised of three mechanisms: fact-finding, explanation-finding, and generalizing. The engineering method is comprised of three analogous mechanisms: function-finding, form-finding, and exapting. Importantly, both methods create new science and technology by standing on the shoulders of existing science and technology. The engineering method draws heavily on science and the sophistication of the models that science enables. The scientific method draws heavily on experimental technology and the sophistication of the experimentation that such technology enables, while the verification of explanation draws heavily on calculational technology and the sophistication of the verification that such technology enables.

Two iconic examples are the transistor (and its associated transistor effect) and the maser / laser.

During the five-year period between 1944 and 1949, three new forms were found for the transistor in three fundamentally different ways (as a thin-film FET, as a dual-point-contact transistor, and as a bipolar junction transistor), each making use of interim scientific ideas as they were evolving in real time. During the same period, those interim scientific ideas also went through three stages: initial ideas for triode-like FET phenomena; then ideas about surface states in response to the unsuccessful thin-film FET experiment; then ideas about minority carriers in response to the dual-point-contact-transistor experiment, whose results did not follow the expectation of field-effect theory.

During the six-year period between 1948 and 1954, the maser was engineered, an exemplar of the use of scientific and engineering insights (stimulated emission, population inversion, molecular beam separation, radio frequency resonators) to find a form for a complex new device. Out of this successful engineering demonstration grew a deeper scientific understanding of the maser, particularly its unexpected frequency purity described by the now-well-established Schawlow-Townes equation. During the subsequent six-year period between 1954 and 1960, the laser was engineered, making use of two key insights: the scientific insight that three-level systems could expand the class of gain media beyond those compatible with molecular beam separation, and the engineering insight to use Fabry-Perot cavities for optical feedback.

2

The Intricate Dance of Question-and-Answer Finding

In Chapter 1, we articulated our first stylized fact, that science and technology are two equally important but qualitatively different repositories of technoscientific knowledge and that these repositories coevolve deeply interactively to create new science and technology. This stylized fact characterizes what one might call the "horizontal" structure of technoscientific knowledge, in the sense that science and technology are coequal repositories of knowledge, neither being above or below the other in any way. We also noted that the central elements of those repositories (for science, facts and their explanations; for technology, functions and the forms that fulfill them) are nested, often deeply.

In Chapter 2, we turn to our second stylized fact: that the nesting of these central elements can be thought of as questions and answers organized into loose hierarchically modular networks, and that these questions and answers coevolve interactively to create new questions and answers. This stylized fact characterizes what one might call the "vertical" structure of technoscientific knowledge. The dynamics that play out in this structure are intuitive to bench researchers: finding new facts or functions is the finding of new questions to answer; finding new explanations for the fact or new forms to fulfill those functions is the finding of new answers to those questions. Question-finding and answer-finding go hand in hand and bolster each other in a symbiotic union. *Both* are important to advancing knowledge. As discussed in the introduction, however, an opposing belief is widespread but mistaken: the belief that the goal of research is to answer questions rather than to also find new questions. In this chapter, we correct that belief.

We begin with the network of questions and answers: how it is organized as a loose modular hierarchy that has been called (and to which we will refer as) a "seamless web" of knowledge (Anderson, 2001). Then, we discuss the mechanisms by which questions and answers coevolve interactively. Just as with science and technology, because the existing body of questions and answers (Q and A) is easy to confuse with the creation of new questions and answers ($\dot{Q}$ and $\dot{A}$), we distinguish these symbolically, with dots above quantities indicating time rates-of-change of those quantities. Finally, we give two concrete examples of the interactive co-evolution of $\dot{Q}$ and $\dot{A}$: the first from science (the theory of special relativity) and the second from technology (the iPhone).

Throughout, we use the phrase "questions and answers" to refer to both scientific and technological knowledge. However, we note that the phrase "questions and answers" is slightly more commonly used for scientific knowledge while the phrase "problems and solutions" is slightly more commonly used for technological knowledge. Here, we take the two phrases to be interchangeable: a problem to be solved can be thought of as a question to be answered, and a solution to a problem can be thought of as an answer to a question. Thus, when we use the phrase "questions and answers," we will often be referring simultaneously to both scientific questions / answers and technological problems / solutions.

2.1 Networks of Questions and Answers: *Q* and *A*

We start with the two networks of questions and answers: one for scientific the other for technological knowledge, each organized as a loose modular hierarchy. The networks are hierarchical in the sense that scientific knowledge (facts and their explanations) and technological knowledge (functions and the forms that fulfill them) are both nested, often deeply. The networks are modular in that closely related questions and answers cluster together to form scientific knowledge domains and technological components. The networks are thus modular hierarchies, but highly cross-connected ones, each forming a self-reinforcing, seamless web of questions that are answered in multiple ways and answers that are reused and repurposed to answer multiple questions. Finally, as the network is traversed up and down its hierarchy, some features are similar, but others different, at different levels in the hierarchy.

The Network Is Hierarchical: The Nesting of Questions and Answers

The way in which scientific and technological knowledge are hierarchical is depicted in Figure 2-1 and stems from the nesting discussed in the last chapter, both of scientific facts and explanations and of technological functions and the forms that fulfill them.

In science, at the top of the hierarchy are facts—raw patterns in observed phenomena. These patterns can be thought of as questions: Why does a particular pattern occur? When one releases a ball, why does the ball fall, and fall faster the farther it has fallen? Explanations of those raw patterns come a level below in the hierarchy, and can be thought of as answers to those questions: Galileo's sixteenth-century explanation of the observed distance-versus-time pattern was that the velocities of falling balls increase linearly with time. But this answer, or explanation, becomes itself another question: Why do the velocities of falling balls increase linearly with time? This question begs a deeper explanation, a deeper answer: Newton's explanation was that gravity is a force, that uniform forces cause uniform acceleration, and that uniform acceleration causes linear increases in velocity. Scientific understanding is

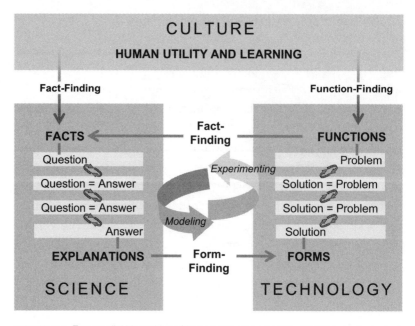

FIGURE 2-1. Two parallel hierarchies of knowledge, one scientific, the other technological.

always incomplete, of course, so there is always a point at which we have no deeper explanation. This in no way detracts from the power of the explanations that do exist: science seeks proximate whys but does not insist on ultimate whys. The general theory of relativity explains Newton's laws of gravity, even if its own origin is yet to be explained.

In technology, at the top of the hierarchy are human-desired functions. These functions present problems that are solved by forms below them in the hierarchy. Forms fulfill functions, but those forms present new problems that must be solved at successively deeper levels. Shifting from the problem-solution nomenclature to the equivalent question-answer nomenclature, we can say that the iPhone represented a technological question: How do we create an Internet-capable cellular phone with a software-programmable interactive display? A partial answer came in the form of multitouch capacitive surfaces, which opened up a significant design space for user interaction when multiple fingers operate simultaneously. But the opaqueness of existing multitouch surfaces itself became a question: How do we make multitouch surfaces transparent so that the display is visible? The multitouch transparent surface display provided an answer.

In other words, science and technology are both organized into hierarchies of question-and-answer pairs, with any question or answer having two "faces." One face, pointing downward in the hierarchy, represents a question to an answer just below it in the hierarchy. The other face, pointing upward in the hierarchy, represents an answer to a question just above it in the hierarchy. We emphasize that our depiction of questions as "above" answers and answers as "below" questions is arbitrary—it does not signify relative importance or value but is simply intended to be consistent with common usage. In science, an explanation is deeper and more "foundational" than the fact it explains, especially if it generalizes to explanations of many other facts. Special relativity is, in that sense, deeper than the constancy of c because it answers the question of why c is constant; it also answers the question of how much energy is released during nuclear fission and fusion. In technology, forms are deeper and more "foundational" than the functions they fulfill, especially if they have been exapted to fulfill many other functions. The multitouch transparent surface display is more foundational than the iPhone because it not only helps answer the question of how to create the iPhone, but also helps answer the question of how to create human-interactive displays in general. Rubber is more foundational than a bicycle tire because it not only helps answer the question of how to create a bicycle tire, but also helps answer the question of how to create a myriad of other kinds of tires.

The Network Is Modular: Facilitating Exploitation and Exploration

The way in which scientific and technological knowledge is modular is depicted in Figure 2-2. Closely related scientific questions and answers are organized into what we might call scientific domains, which we will refer to as scientific knowledge modules. Closely interacting technological problems and solutions are organized into engineered components, which we will refer to as "technological knowledge modules."

Closely related scientific questions are often answerable within a scientific knowledge domain, or "scientific knowledge module," drawing on multiple subdomains nested within the larger domain. A question related to some electron transport phenomenon in a particular semiconductor structure lies in the broad domain of semiconductor science

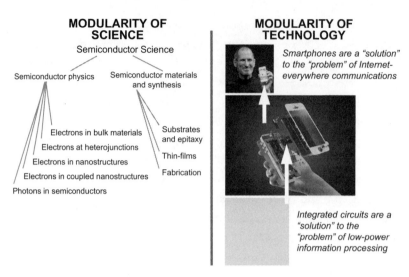

MODULARITY OF SCIENCE

Semiconductor Science

Semiconductor physics

Semiconductor materials and synthesis

Electrons in bulk materials
Electrons at heterojunctions
Electrons in nanostructures
Electrons in coupled nanostructures
Photons in semiconductors

Substrates and epitaxy
Thin-films
Fabrication

MODULARITY OF TECHNOLOGY

Smartphones are a "solution" to the "problem" of Internet-everywhere communications

Integrated circuits are a "solution" to the "problem" of low-power information processing

FIGURE 2-2. Modularity of scientific and technological knowledge. *Credit:* Photo of Steve Jobs with iPhone, Matthew Yohe / Wikimedia Commons. Exploded view of iPhone courtesy of Rajkumar Remanan. Graphic of integrated circuit, David Carron / Wikimedia Commons.

but the answer might require an integrated understanding of both the subdomain of electron transport physics as well as the subdomain of the materials science of the synthesized structure. The subquestion associated with electron transport physics might require an integrated understanding of the subdomain of electrons in various kinds of structures (bulk materials, heterojunctions, nanostructures, coupled nanostructures) and of the sub-subdomains of interactions of electrons with phonons in those structures. The subquestion associated with the materials science of the synthesized structure might require an understanding of the sub-subdomains of substrates and epitaxy, thin films, or post materials synthesis fabrication. In other words, we can think of scientific knowledge domains as a modular hierarchy, and think of its subdomains as submodules and sub-submodules.

Closely related technological problems, likewise, are often solved by key technological components (technological knowledge) that may nest multiple subcomponents within the larger components. An iPhone is a component that is itself composed of many subcomponents, and each subcomponent is similarly subdivided. We can think of the "problem" of the iPhone as a component as that is "solved" by its subcomponents—an enclosure, a display, a printed circuit board, a camera, and input / output ports. We can think of the "problem" of a printed circuit board as a subcomponent that is "solved" by sub-subcomponents that include low-power integrated circuit chips. Conversely, an iPhone is also a component that is itself nested in a hierarchy of use functions. An iPhone might be used as a solution to the problem of "running" a text-messaging app; a text-messaging app might be used as a solution to the problem of sending a mass text message to a friend group; the mass text message might be used as a solution to the problem of organizing the friend group into a protest in Times Square; and the protest in Times Square might be part of a solution to the problem of organizing a wider social movement for some human-desired social cause.

One might ask: Why is scientific and technological knowledge modular? They are modular because they are complex adaptive systems (systems sustained by and adapted to their environment by complex internal changes), and virtually all complex adaptive systems are modular (Simon, 1962). Complex adaptive systems both exploit their environments and explore their environments to improve that exploitation. Modularity enables efficiency, both in the exploitation of existing knowledge about the environment and in the exploration of that environment to create new knowledge.

Modularity and Exploitation

How does modularity of knowledge enable more efficient *exploitation* of that knowledge? In two ways.

First, single human brains have finite cognitive ability and cannot know all that is useful to know about our vast world. Specialization and division of labor is necessary. For example, the explanation of faint patterns in the Laser Interferometer Gravitational-Wave Observatory benefits from specialists in general relativity, statistics, astrophysics, laser interferometry, and so on—no generalist could provide explanations with as much confidence as a collection of specialists. The building of an optical fiber communications system benefits from specialists in laser diode sources, high-speed modulation, optics, glass fibers, detectors, and communication protocols. Indeed, as emphasized by economists since Adam Smith, the importance of specialization and division of labor increases with the extent of the market. Human knowledge is no exception, as articulated by the economist and complexity theorist Brian Arthur (Arthur, 2009, p. 37):

> Modularity, we can say, is to a technological economy what the division of labor is to a manufacturing one; it increases as given technologies are used more, and therefore as the economy expands. Or to express the same thing in good Smithian language, the partition of technologies increases with the extent of the market.

With the human population in the billions, the market for human knowledge is huge. In such an environment, the degree of specialization and division of labor is extremely pronounced.

Second, modularity reflects the way the world actually is—a world of space and time in which large slow-moving things (for example, wind turbines) are composed of smaller faster-moving things (atoms and electrons), with different phenomena governing behaviors at the different length and time scales. In a world built according to a plan characterized by such decomposability, scientific and technological knowledge about that world will reflect that decomposability.

With respect to science and its facts and explanations, psychologist and economist Herbert Simon would say that (Simon, 2001, p. 56):

> In a world built on this plan, as large aspects of our world are, phenomena on different time scales can be studied nearly independently of each other; and a good deal of the specialization among the sciences is based precisely on this possibility for decomposition and aggregation. Each of the sciences selects phenomena on a particular time scale, treats variables on slower time scales as constant parameters, and treats subsystems operating on faster time scales as aggregates whose internal details are irrelevant at the scale of special interest. This partitioning is perhaps seen in its sharpest form in physics and chemistry, in the progression from quarks, through the so-called elementary particles, atomic components, to small molecules, macro-molecules, etc.

An example of this is the Born-Oppenheimer approximation in quantum chemistry and molecular physics. Nuclei, composed of neutrons and protons, are heavy and move slowly, hence they can be considered fixed in space when we solve for the dynamics and energetics of the electrons, which are light, move quickly, and "glue" the nuclei together into molecules.

With respect to technology and its functions and forms, there is a similar matching of spatial and time scales, and of phenomena. Transporting large and heavy objects benefits from large mechanical technologies; transporting small and light objects benefits from small mechanical technologies. Transporting electrons with charge makes use of voltages and electric fields; transporting photons makes use of optical refractive index variations. Just as with scientific explanations of the world, technological forms for interacting with the world are modular, in part because the world itself is modular.

Modularity and Exploration

How does modularity of knowledge enable more efficient *exploration* that creates new knowledge? It does so by enabling parallel evolution of submodules without compromising the function of the larger module. Parallel evolution, in turn, is possible because of *protocols*: the ways in which scientific and technological knowledge modules interact with each other. In science, the equation $E = mc^2$ (a relationship between energy and mass) is a compact way in which the theory of special relativity interacts with and informs nuclear chemistry, the subfield of chemistry dealing with nuclear reactions. Explanations for why $E = mc^2$ might change, but so long as $E = mc^2$ itself is preserved, nuclear chemistry will not need to change. In technology, Ohm's law $V = IR$ (the relationship between voltages and currents) is a compact way in which resistors interact with the circuits in which they are embedded. The materials from which resistors are fabricated might differ, but so long as $V = IR$ is preserved, the circuits in which they are embedded will not need to change.

In other words, as long as a particular module respects the protocols by which it interacts with other modules, it is free to provide explanations, or forms, in different, possibly improved, ways. As Herbert Simon, using an example from evolutionary biology would explain (Simon, 2001, p. 58):

> The liver converts ammonia previously derived from amino acids into urea, which it delivers to the kidneys for excretion. Provided that the urea is synthesized and delivered, the operation

of the kidneys is unaffected by the exact sequence of urea-synthesizing reactions in the liver. Increased efficiency of the liver will not affect the kidneys in the short run; over a longer run the ratio of sizes of the two organs may change if one gains efficiency more rapidly than the other. The important fact is that the rate of evolution is accelerated by the mutual independence of subsystems from sensitivity to each other's details of structure and process.

So long as module improvement respects the protocols by which the module connects to other modules, module improvement can proceed independently of those other modules.

An extreme case of this is when the protocols are between different levels of the modular hierarchy and when there is richness on both sides of the protocol. When the upper side of the protocol is rich, the knowledge base on the lower side of the protocol is often referred to as a "platform" on which knowledge modules above it can be based. In science, Newton's laws were a platform on which both celestial and terrestrial mechanics could be based. In technology, the personal computer software operating system is a platform on which a rich set of software applications can be based. Moreover, when the lower side of the protocol is also rich, the shape of the knowledge network becomes hourglass-like (Saltzer et al., 1984). In the case of technological knowledge, the waist of the hourglass is a distinguished layer or protocol, with technologies underneath implementing the protocol and technologies above building on the protocol—with both sides "screened" from each other by the protocol itself. As a result, the number of applications explodes independent of implementation details; similarly, the number of implementations explodes independent of application details. The number of software applications built on the Windows operating system is enormous; the number of hardware and software implementations of the Windows operating system is also enormous.

In other words, imagine two complex adaptive systems, one organized modularly and one not. At one moment, both might be able to exploit their environments equally and thus be equally "adapted" to their environment. But they will evolve at vastly different rates, with the one organized modularly quickly outstripping the one not so organized

(Simon, 1962). Modularity appears to be an evolved property in biology, one that is mimicked in the organization of human knowledge.

The Network Is *Networked:* A Self-Reinforcing "Seamless Web"

Human technical knowledge (and the question-and-answer pairs that comprise that knowledge) are modular, but only very loosely modular.

In science, explanations of one set of facts generalize to explanations of other sets of facts, thereby creating a large-scale interconnectedness of facts and explanations across the network of questions and answers. As articulated by Phil Anderson, winner of the 1977 Nobel Prize in Physics (Anderson, 2001, p. 488):

> the logical structure of modern scientific knowledge is not an evolutionary tree or a pyramid but a multiply connected web. . . . The failure to recognize this interconnectedness becomes obvious when we are presented with 'classical Newtonian mechanics, quantum mechanics, quantum field theory, quantum electrodynamics, Maxwell's electromagnetic theory' and, in a separate place, 'fluid dynamics,' as logically independent and separate rather than as, what they are, different aspects of the same physical theory, the deep interconnections among them long since solidly cemented.

In technology, forms are composed of subforms in a fanout of linkages moving downward in the hierarchy; but subforms are used and reused, contributing to many other modules in a fanout of linkages moving upward. As a form, the iPhone depends on subforms (for example, semiconductor memory chips) that, themselves, contribute to many different forms, including mainframe computers.

In other words, knowledge modules at one level link, not just vertically (to modules at higher or lower levels in the knowledge hierarchy), but also cross-link diagonally to other modules. Thus, science and technology each form self-reinforcing "seamless webs" of knowledge (Anderson, 2001). The reuse and repurposing of questions and answers ("generalization" in science and "exaptation" in technology) are powerful mechanisms for internal self-consistency and for explanatory and functional power. Neither science nor technology exists as a vulnerable "house of cards." Instead, the "cards" are resistant to falling because they are structurally interconnected. The power of these tight cross connections is similar to the power of "small world" networks, whose nodes are both clustered into modules but also "weakly linked" to distant other nodes (Buchanan, 2003).

Of course, the very interlocking nature of the networks of scientific and technological knowledge also means that there is rigidity. Each connection from one module to another represents a protocol to be respected. A new module might in principle introduce huge improvements if it were to replace an existing module in the network. But if the new module requires different ways of interacting with the multiplicity of its interconnecting modules, then it will be disfavored. Instead, the current module that respects current protocols will be favored, or "locked in." Achieving the benefits of the new module may require concerted changes in its interconnecting modules, and the more interconnections, the greater chance other modules will also need to change. When an explanation has been found, not only for one but for a multiplicity of facts, the explanation becomes embedded holistically into a larger web of internally consistent scientific knowledge, and, as again articulated by Phil Anderson (Anderson, 2011, p. 138):

> it becomes much harder to modify any feature of it without "tearing the web"—[without] falsifying a proliferating network of verified truths. It is no longer possible to falsify evolution for instance, without abandoning most of physics and astronomy as well, not to mention the damage that would be done to the many cross-links among paleontology, plate tectonics, physics, and planetary astronomy.

Likewise, when a technological form helps fulfill a multiplicity of functions, the form becomes embedded in a larger web of internally self-reinforcing technological knowledge, with significant inertia acting against its replacement. The deep embeddedness of von Neumann computation architectures in virtually all computing applications delayed significantly the use, beginning in the 2010s, of the graphical processing units that have now revolutionized deep neural network computation (Hooker, 2020).

Such lock-in can also be exacerbated in a number of ways. If the modules are "lower" in the vertical hierarchy of knowledge, the higher-level modules that "depend" on that module and will resist its change. The social networks that evolve alongside knowledge networks are a conservative social force that resists change, as famously articulated by Max Planck (Planck, 1950, pp. 33–34):

> A new scientific truth does not triumph by convincing its opponents and making them see the light, but rather because its opponents eventually die, and a new generation grows up that is familiar with it.

Finally, the public-sector, by subsidizing demand in an attempt to accelerate the advance of a technology perceived as having societal importance, can cause premature increases in manufacturing scale that lock in immature and inferior versions of the technology (Funk, 2013).

Similarity Up and Down the Hierarchy: More Is Different

The knowledge hierarchy appears to be self-similar in that every level of the hierarchy resembles every other level of the hierarchy in having upward- and downward-facing faces: question-and-answer pairs all the way up and down the hierarchy. Nevertheless, there is an asymmetry. Down is not the same as up, and knowledge deeper in the hierarchy might be thought of as more important than knowledge shallower in the hierarchy. An extreme view might be that shallower knowledge is derivable from (or composed of) deeper knowledge by reductionism, and so shallower knowledge is less important and more trivial. In this way of thinking, the particle physicist might deservedly be arrogant, as was the discoverer of the positron, Carl Anderson, when he said, "The rest is chemistry"; or as was the discoverer of atomic nuclei, Ernest Rutherford, when he said, "All science is either physics or stamp collecting" (Birks & Rutherford, 1962, p. 108). Likewise, the molecular biologist might deservedly be arrogant, determined to try to reduce everything about the human organism to "only" chemistry, from the common cold to mental disease to the religious instinct. It can be but a short step to the ranking of knowledge that was given by the philosopher August Comte: from mathematics to astronomy to physics to chemistry to biology to sociology—with the latter considered derivative of the former and thus more trivial.

But this way of thinking is manifestly incorrect. Knowledge situated higher in the hierarchy is no more uniquely derivable from knowledge lower in the hierarchy than the components of nature are uniquely derivable from their subcomponents. To expand on this for the case of technological knowledge, it is certainly true that subforms are restricted in the ways they can be put together into higher level forms. But in the vast space of possibilities for what forms to create, which of those possibilities actually emerge cannot be derived solely from knowledge of just the subforms themselves.

How, then, is the space of possibilities narrowed?

The first way in which the space of possible forms is narrowed is function. Subforms are versatile and can be put together in many different ways, but only those ways that create forms that fulfill desired function will be selected for—and function is dictated from above, not below, in the knowledge hierarchy. Indeed, the importance of function in selecting form is illustrated by the degree to which functions have been fulfilled by sometimes wildly different forms. Countless forms for fulfilling video display functionality (cathode-ray tubes, plasma panels, liquid crystals, micromirrors, organic light-emitting diodes, inorganic micro light-emitting diodes) have been created since the invention of the television in 1927.

The second way in which the space of possible forms is narrowed is historical contingency—the specific sociocultural technoscientific sequence of events culminating in the form (Jacob, 1977). As one goes up the knowledge hierarchy, from elementary particles to atoms to molecules to human beings, phenomena become less symmetric and more complex, with inherent ambiguity, noninevitability, and historical contingency in the precise ways in which the symmetries mentioned above are broken and complexity is introduced. Electroweak symmetry could conceivably have been broken in different ways in the early stages of the universe, but the way it was actually broken gave us the super-heavy Higgs boson and the massless photon. Right-handed molecules versus left-handed molecules are both "allowed" by the reductionist rules one level down, but left-handed molecules predominate in life on earth (though they might not in an alien civilization we might find in another galaxy). Similarly, many data communications protocols are possible, but TCP / IP (Transmission Control Protocol / Internet Protocol) predominates.

Put another way, "reductionism" does not imply "constructivism." Just because a form can be reduced to subforms *a posteriori* does not mean that it can be uniquely constructed from those subforms *a priori*. As articulated by the physicist Phil Anderson (Anderson, 1972, p. 393):

> The main fallacy in the extreme reductionist point of view is that the reductinist hypothesis does not by any means imply a "constructionist" one: The ability to reduce everything to simple fundamental laws does not imply the ability to start from those laws and reconstruct the universe. Just because Y lies below X in the hierarchy does not imply that science X is "just applied Y." At each stage entirely new laws, concepts, and generalizations are necessary, requiring inspiration and creativity to just as great a degree as in the previous one. Psychology is not applied biology, nor is biology applied chemistry.

Indeed, it is precisely because the narrowing of constructivist possibilities does not solely derive from reductionist knowledge that, in science, observation and experiment are so necessary. As articulated by Robert Laughlin and David Pines (Laughlin & Pines, 2000, p. 30):

> For the biologist, evolution and emergence are part of daily life. For many physicists, on the other hand, the transition from a reductionist approach may not be easy, but should, in the long run, prove highly satisfying. Living with emergence means, among other things, focusing on what experiment tells us about candidate scenarios for the way a given system might behave before attempting to explore the consequences of any specific model. This contrasts sharply with the imperative of reductionism, which requires us never to use experiment, as its objective is to construct a deductive path from the ultimate equations to the experiment without cheating. But this is unreasonable when the behavior in question is emergent, for the higher organizing principles—the core physical ideas on which

the model is based—would have to be deduced from the under-lying equations, and this is, in general, impossible.

Thus, the importance of, and the challenges to advancing, knowledge does *not* depend on its level in the hierarchy. From quantum electrodynamics to fluid dynamics to DNA biochemistry to human ethology, at every level of the knowledge hierarchy are entirely new and different conceptual structures, none superior or inferior to those at other levels. To state, for example, that the hydrogen atom H is somehow superior to the dihydrogen molecule H_2, or that biology is somehow inferior to chemistry, which in turn is somehow inferior to physics, is nonsensical. Knowledge at every level of the knowledge hierarchy has a distinct "life" constrained and enriched by knowledge in which it is embedded at deeper *and* shallower levels. To again quote Phil Anderson (Anderson, 1972, p. 396):

> We expect to encounter fascinating and, I believe, very fundamental questions at each stage in fitting together less complicated pieces into the more complicated system and understanding the basically new types of behavior which can result.

Dissimilarity Up and Down the Hierarchy: More Is Different

We just discussed the self-similarity of knowledge up and down the hierarchy. Knowledge is equally important and nontrivial to create throughout the hierarchy. There are limits, however, to that self-similarity. Knowledge does differ, depending on how deep or shallow in the hierarchy it is.

A first dissimilarity in knowledge at various levels of their hierarchies is the degree of dependency on knowledge at other levels of the hierarchies. The lower, or deeper, knowledge is in the hierarchy, the more knowledge lies above it and is dependent on, it. Knowledge production lower in the knowledge hierarchy thus has a greater potential for human knowledge transformation throughout the hierarchy above but realizes that potential with greater difficulty due to network lock-in effects.

A second dissimilarity in knowledge at various levels of their hierarchies is the degree to which knowledge becomes more historically contingent and complex the higher it is in the hierarchy. Historical contingency and complexity are both cumulative: the higher that knowledge is in the hierarchy, the greater the accumulation of historical contingency from the layers below and the greater the number of knowledge submodules of which it is "composed." Moreover, as knowledge becomes more historically contingent and complex, the more empirical and tacit (the less codifiable via simplifying general principles) it becomes. The more tacit the knowledge, the less duplicable and more easily kept secret; and the more easily kept secret, the more monetizable. An equation like $E = mc^2$ is easily codified and difficult to keep secret, but insights into situation-specific human behaviors are less easily codified and easier to keep secret. Thus, on the one hand, the shallower in the hierarchy the newly created knowledge, the greater the incentive for the for-profit sector to support and then monetize it. On the other hand, the deeper in the hierarchy the newly created knowledge, the less easy to monetize—leading to the well-known failure of markets to support research and the need for the not-for-profit sector (government agencies and philanthropies) to support it. Perhaps in part because of this, knowledge production higher and lower in the hierarchy tends to be given different names. Lower, or deeper, in the hierarchy, knowledge production tends to be called "research and development," while higher in the hierarchy, knowledge production tends to be called "innovation." Interestingly, the tendency for knowledge to be more tacit and less codified applies not only to knowledge higher in the hierarchy but also to knowledge that is more technological and less scientific in nature. Thus, technological knowledge, like knowledge shallower in the hierarchies, is more easily monetized, again with significant consequences for how research is supported.

2.2 Finding New Questions and Answers: $\dot{Q}$ and $\dot{A}$

Questions and answers, as just discussed, are the fundamental way in which knowledge is networked and organized. Viewed as *static* repositories of knowledge, the two question-and-answer networks, one for science (S) and one for technology (T), are *not* connected to each other. Scientific explanations are not the same as technological forms. Explanations rest logically on deeper explanations just as forms are composed of more basic forms, but explanations do not rest logically on forms, just as forms are not composed of explanations.

Not so for the *dynamical* evolution of the networks to create new science ($\dot{S}$ and $\dot{T}$), however. In the science network, the finding of new questions and answers depends on questions and answers in *both* the science and technology networks; just as the finding of new questions and answers in the technology network depends on questions and answers in *both* the science and technology networks. Both $\dot{Q}$ and $\dot{A}$ "stand on the shoulders" of existing Q and A throughout the science and technology networks. Here, we discuss various mechanisms that mediate $\dot{Q}$ and $\dot{A}$. Elaborating on concepts from Stuart Kauffman, Steven Johnson, and others (Johnson, 2011; Kauffman, 1996), we distinguish the mechanisms according to how distantly they explore beyond the current body of questions and answers. If the questions and answers being explored already exist in the here and now, then the exploring happens in the realm of the "possible." If, however, they do not exist in the here and now but are latent and come to exist through new recombinations of existing questions and answers, then the exploring is in the "adjacent possible." If the questions and answers being explored exist in neither the possible nor the adjacent possible but are two or more levels removed from the possible, then the exploring is in the "next-adjacent possible."

The Possible

The "possible" is where existing question-and-answer pairs lie. Just because the pair already exists, however, does not mean there is no room for improvement. Any question-and-answer pair comprises many "parts" (the question being answered by a combination of knowledge modules), where continuous adjustments to improve the "fit" between parts is almost always possible.

Starting with a question, we look downward in the knowledge hierarchy to readjust our answers. An integrated circuit, for example, might be decomposed into transistors, resistors, and capacitors, which might themselves be decomposed into various semiconductor, oxide, and metal materials. The transistor might be improved through engineered variations in the thickness or smoothness of the various materials. The integrated circuit might be improved directly from these improvements in the transistor, and it might be improved further if its architecture and layout are modified so as to better take advantage of interactions between the improved transistor and its neighboring resistors, and capacitors. Maybe the improved transistor requires less drive current, so the resistors can be changed to accommodate. Maybe the improved transistor requires less total electronic charge to switch, so the capacitors can be changed to accommodate. It is rare that a question-and-answer pair is fully optimized; it can almost always be further optimized through myriad continuous changes in its various components and their connections. And, even if it is locally and temporarily optimized, some exogenous change might occur: a component that plays dual use in answering a different question might change as it is optimized for that other question. All such changes create room for continuing optimization.

Starting with an answer, we look upward in the knowledge hierarchy to readjust our question. Consider the common practice of A / B testing. A / B testing is a way of getting at what one's customer prefers, a way of refining questions that are already being answered to some degree by a technology. In the context of web pages, where A / B testing was first developed in 2000 by Google (Wikipedia contributors, 2019), two web page variants (A and B) are developed and presented to users. The users' responses define which variant users prefer (which variant reflects the question that users want answered), and that variant is then selected. It can be thought of as "evidence-based practice," except the evidence is associated, not with scientific truth, but rather with user preferences: what questions should the technology try to answer, what user problems should the technology try to solve, and what functions should the technology try to fulfill. Because A / B testing resides in the "possible," though, it has limitations. As articulated by Ken Kocienda (Kocienda, 2018, p. 212):

> In this kind of test, commonly referred to in the high-tech industry as an A / B test, the choices are already laid out. In this Google pick-a-blue experiment, the result was always going to be one of those forty-one options. While the A / B test might be a good way to find the single most clickable shade of blue, the dynamic range between best and worst isn't that much. More important, the opportunity cost of running all the trials meant there was less time available for everyone on the development

team to dream up a design that people might like two, or three, or ten times more. A / B tests might be useful in finding a color that will get people to click a link more often, but it can't produce a product that feels like a pleasing and integrated whole. There aren't any refined-like responses, and there's no recognition of the need to balance out the choices. Google factored out taste from its design process.

Nonetheless, it is a powerful way of reconfiguring and optimizing the questions that a technology should try to answer. Indeed, the degree to which A / B testing permeates user-interface development is an indication of how difficult it can be to know what questions users want answered and of how difficult question readjustment can be.

The Adjacent Possible

The "adjacent possible" is where latent questions and answers (those yet to be found, but only one level removed from the actual) exist. As articulated elegantly by Steven Johnson (Johnson, 2011, p. 30):

> The scientist Stuart Kauffman has a suggestive name for the set of all those first-order combinations: "the adjacent possible." The phrase captures both the limits and the creative potential of change and innovation. In the case of prebiotic chemistry, the adjacent possible defines all those molecular reactions that were directly achievable in the primordial soup. Sunflowers and mosquitoes and brains exist outside that circle of possibility. The adjacent possible is a kind of shadow future, hovering on the edges of the present state of things, a map of all the ways in which the present can reinvent itself. Yet it is not an infinite space, or a totally open playing field. The number of potential first-order reactions is vast, but it is a finite number, and it excludes most of the forms that now populate the biosphere. What the adjacent possible tells us is that at any moment the world is capable of extraordinary change, but only certain changes can happen.

Questions and answers in the adjacent possible result from discontinuous, chance recombinations of knowledge modules that exist in the possible but are not yet linked. As articulated by the creativity psychologist Dean Simonton with respect to the "chance" aspect (Simonton, 2004, pp. 29–30):

> In particular, creativity is often said to be combinatorial—that is, it entails the generation of chance combinations. For instance, Jacques Hadamard (1945), the mathematician, claimed that mathematical creativity requires the discovery of unusual but fruitful combinations of ideas. To find such combinations, it is "necessary to construct the very numerous possible combinations, among which the useful ones are to be found." But "it cannot be avoided that this first operation takes place, to a certain extent, at random, so that the role of chance is hardly doubtful in this first step of the mental process."

Though chance plays an important role, nonetheless when ideas are ripe to be recombined, they often are. Given an entire human society of minds, the chances that more than one mind will take advantage of particularly useful chance recombinations can be high. The adjacent possible is thus where innovation "ripe to happen" happens. One can see this in the phenomenon of the "multiple"—when something in the adjacent possible, having just come into view for one innovator, also comes into view for other innovators at approximately the same time. A few examples of such multiples are the theory of evolution (Charles Darwin and Alfred Wallace), the maser (Charles Townes, Alexander Prokhorov, Nikolay Basov, and Joseph Weber), and the telephone (Alexander Graham Bell and Elisha Gray).

Importantly, the adjacent possible is the home of both latent questions *and* latent answers—innovation can start from either end. It may be that there are existing knowledge modules that would solve a problem that already has a solution but solve it much better (as when fiber optics supplanted electrical wire for long-distance voice communications). It may be that there are existing knowledge modules that would solve a problem that has not yet been posed but is ripe to be posed (as in 3M's Post-it Notes solving the problem of temporary tags attached to pages in documents). There are thus two faces to the exploration of the adjacent possible: a downward face (answer-finding) and an upward face (question-finding).

The downward face, answer-finding, starts with a question or problem and seeks an answer or solution. In economics this face represents demand pull: a human or market need or demand that drives the finding of a way to meet it. In science, this face represents explanation-finding: finding an explanation for some observed pattern. In technology, this face represents form-finding—finding a form that fulfills a desired function. As will be discussed in Chapter 4, this face also represents the vast majority of modern, formal research, in which questions and problems are known but answers and solutions are not. Indeed, it is not uncommon for experts in a field to gather in advisory committees to define, from the top down, outstanding "challenges" in their field; these grand challenges lead to a solicitation for proposed answers and the proposed answers then lead to projects executed with the hope of finding answers that do indeed solve those grand challenges.

Answer-finding begins with reductionism, reducing problems into constituent functionalities and analyzing whether those constituent functionalities can be met or solved, and ends with integration, integrating the constituent functionalities into a synthesized solution. Paraphrasing the Greek mathematician Pappus (300 BCE), the Hungarian mathematician George Polya articulates it thus (Polya, 2014, p. 141):

> In analysis, we start from what is required, we take it for granted, and we draw consequences from it, and consequences from the consequences, till we reach a point that we can use as a starting point in synthesis. . . . We inquire from what antecedent the desired result could be derived; then we inquire again what could be the antecedent of that antecedent, and so on, until passing from antecedent to antecedent, we come eventually upon something already known or admittedly true. . . . But in synthesis, reversing the process, we start from the point which we reached last of all in the analysis, from the thing already known or admittedly true. We derive from it what preceded it in the analysis, and go on making derivations until, retracing our steps, we finally succeed in arriving at what is required.

In a sense, the final integration, or synthesis, can to some extent be thought of as simply a check for the correctness of the reductionist analysis. If the synthesis does not "work," then one goes back to the hard work of the analysis. But this is by no means to suggest that the synthesis is easy—the actual building of a technology from constituent subtechnologies often presents its own difficult problems requiring its own ingenious solutions.

The upward face, question-finding, starts with answers or solutions to some questions or problems and seeks a *new* question or problem to which those answers / solutions (or combinations of answers / solutions), might apply. In economics, this face represents supply push rather than demand pull. Supply push offers a way of doing something that might satisfy a not-yet-met market demand and that drives development of that demand ("build it and they will come"). In science, this face represents generalization: an explanation for an observed pattern in some phenomenon that also explains an observed pattern in some other phenomenon. In technology, this face represents exaptation: when a form that fulfills some function also fulfills another. Despite the importance of generalization and exaptation, they are seldom emphasized in contemporary, formal research that prefers to answer questions already known. Proposals are certainly easier to write when one already knows the question and has some ideas about how to answer it.

Question-finding begins with synthesis, taking answers or solutions that already exist and integrating them to discover new questions they might answer or problems they might solve. Then, once a new question has been tentatively found, it must be followed by reductionist analysis to verify that the answer / solution does indeed answer / solve the question or problem. Galileo's answer to the question of why balls rolling down an incline will roll back up to their original heights is co-opted by Christiaan Huygens to solve the question of why pendula released at any point in the swing rise to the symmetric point on the opposing swing. For Huygens's question, Galileo's answer was indeed applicable, but for other questions it may not be, which is why synthesis and analysis are both important. Synthesis must be followed by analysis to weed out bad guesses. Indeed, this is one reason why outstanding research is so difficult; researchers rarely have the temperament for both synthesis and analysis. Interestingly, because question-finding looks upward in the knowledge network, it cannot draw, the way answer-finding can, on a systematic reductionist and analytic hunt for answers deeper within a discipline. The hunt for questions typically lies outside the discipline, hence it depends more on serendipity and chance collisions of ideas.

Whether we are question- or answer-finding as we explore the adjacent possible, our search depends on what is present in the possible—on the knowledge modules available for recombination, the knowledge we have at our fingertips.

Successful answer-finding is contingent, in science, on knowing a library of facts and explanations and, in technology, of forms and functionalities. Finding a y that satisfies the equation $y^2 - 13y + 36 = 0$ is trivial, but only if one is familiar with quadratic equations. As articulated by Brian Arthur regarding the invention of the cyclotron, for which Ernest Lawrence won the 1939 Nobel Prize in Physics (Arthur, 2009, p. 121):

> In retrospect, Lawrence's insight looks brilliant, but this is largely because the functionalities he uses are not familiar to us. In principle, Lawrence's problem does not differ from the mundane ones we handle in daily life. If I need to get to work when my car is in the repair shop, I might think: I could take the train and from there get a cab; or I could call a friend and get a ride if I were willing to go in early; or I could work from home providing I can clear up some space in my den. I am reaching into my store of everyday functionalities, selecting some to combine, and looking at the subproblems each "solution" brings up. Such reasoning is not a mystery when we see it applied to everyday problems, and the reasoning in invention is not different. It may take place in territory unfamiliar to us, but it takes place in territory perfectly familiar to the inventor.

Successful question-finding is contingent on having, at one's fingertips, a library of questions that would be interesting to answer but have not yet been answered, or problems that would be useful to solve but have yet no solution. The existence of such libraries of questions and answers exemplifies Louis Pasteur's "prepared mind." Recall how Pasteur, who studied fermentation hoping to answer the question of how to better produce wine, found that fermentation is a powerful tool for studying chemical transformations more broadly. Or consider Steve Jobs, who saw a "mouse" at Xerox PARC and quickly realized its potential importance for a personal computing user interface.

It is hard work to master such libraries, but innovators must expend the necessary upfront effort. As observed by Brian Arthur in the context of technology (Arthur, 2009, p. 123):

> Originators, however, do not merely master functionalities and use them once and finally in their great creation. What always precedes invention is a lengthy period of accumulating functionalities and of experimenting with them on small problems as five-finger exercises. Often in this period of working with functionalities you can see hints of what originators will use. Five years before his revelation, Charles Townes had argued in a memo that microwave radio "has now been extended to such short wavelengths that it overlaps a region rich in molecular resonances, where quantum mechanical theory and spectroscopic techniques can provide aids to radio engineering." Molecular resonance was exactly what he would use to invent the maser.

The Next-Adjacent Possible

As we have seen, latent questions and answers exist in the adjacent possible, one step removed from the possible. Even further from the possible are ideas in what we might call the "next-adjacent possible" where latent questions and answers exist *two or more* steps removed from the actual. It is where a latent question or answer cannot be found unless another latent question or answer is found first. It is when we find a new question that has no match to any existing answers, or we find a new answer that does not match any existing questions—both requiring something more to come into being first.

In a sense, exploring the next-adjacent possible is simply thinking more than one step ahead—thinking "strategically." In the Land of Oz, if we keep going on the yellow brick road, we might find the Wizard, but first we have to get past the Wicked Witch. Answer-finding in the next-adjacent possible is the breaking apart of a problem, then noticing a key auxiliary problem that, if solved, would enable solving the original problem. And this strategy of breaking problems apart, solving auxiliary problems that arise, then rolling them back up again into an overall solution can continue recursively. As articulated by the Hungarian mathematician George Polya (Polya, 2014, p. 145):

> A primitive man wishes to cross a creek; but he cannot do so in the usual way because the water has risen overnight. Thus, the crossing becomes the object of a problem; "crossing the creek" is the x of this primitive problem. The man may recall that he has crossed some other creek by walking along a fallen tree. He looks around for a suitable fallen tree which becomes his new unknown, his y. He cannot find any suitable tree but there are plenty of trees standing along the creek; he wishes that one of them would fall. Could he make a tree fall across the creek? There is a great idea and there is a new unknown; by what means could he tilt the tree over the creek?

Many of the greatest discoveries and inventions, including special relativity and the iPhone, have this "thinking many steps ahead" flavor.

The Next-Adjacent Possible and Interdisciplinary Thinking

Auxiliary problems are not black-and-white; they are gray. If two things happen to fit "as is," then there is no auxiliary problem, and one stays within the adjacent possible. However, if two things might fit, but "not as is," then one has the auxiliary problem of making them fit, and one has entered the next-adjacent possible. Interestingly, this fitting "as is" or "not as is" maps onto common usage of the terms "multidisciplinary" and "interdisciplinary."

On the one hand, if knowledge *is* combined "as is," then we might call the combination "multidisciplinary." In technology, it might be that a flashlight is combined with a bicycle to enable riding in the dark, for which very little modification of either flashlight or bicycle is required, other than a strap and a bolt. In science, it might be that a theory for electrical heating of a filament is combined with a theory for blackbody radiation to create a composite theory of the radiation emitted when a filament draws electrical current: very little modification of either theory is required other than to "hand off" a single parameter, temperature, from one to the other. In other words, in a multidisciplinary team, individuals might have deep knowledge of their respective domains, but have little need to share that deep knowledge. Each individual plays a part in the combination, and large teams with strong leadership can efficiently coordinate this multidisciplinary combination of diverse knowledge.

On the other hand, if knowledge is *not* combined "as is," then the combination might be called "interdisciplinary" (or "transdisciplinary," an even deeper union). In technology, one might combine a higher-resolution display with a smartphone—but to preserve battery life, other parts of the smartphone must change significantly. In science, one might combine a theory of electricity with a theory of magnetism to create an overarching theory of electromagnetism—requiring significant changes in both theories. In other words, in an interdisciplinary team, individuals not only must have deep knowledge of their own domains, but they must also share that knowledge with other individuals to understand how knowledge in one or more domains must be changed to accommodate its interdisciplinary combination.

Here one finds the commonly experienced "breadth versus depth" intellectual trade-off (Avina et al., 2018). The greater the breadth of knowledge being combined, the more novel, unexpected, and "creative" the resulting new knowledge. Newton combined new concepts of mass, force, and acceleration with his new mathematics of calculus to create a new theory of mechanics. Charles Darwin combined Thomas Malthus's population economics with plant and animal breeding to create the new discipline of evolutionary biology. Robert Noyce and Jack Kilby combined the transistor with planar processing and hybrid integration to create the new discipline of integrated circuits. John Von Neumann combined digital hardware with digital software to create the Von Neumann computer. The greater the breadth of knowledge being combined, however, the more likely the knowledge will not mesh, at least not exactly, and the less likely a bricolage of the combined knowledge will work "as is," and the greater the depth of knowledge required to adjust the knowledge so that the new combination will "work." In other words, the further separated the disciplines from which ideas are drawn, the more likely the recombination will fail—disciplines are disciplines in part because the ideas they contain "fit" with each other. Occasionally, however, the recombination might succeed, even if, at a superficial glance, the recombination might seem to fail. For example, the new high-resolution display requires too much battery power, so it seems incompatible with the smartphone; but deeper knowledge of the smartphone might reveal that battery power being drawn from elsewhere in the phone could be reduced. Or the theory of electricity might seem incompatible with the theory of magnetism, but deeper knowledge of the empirical facts associated with each would reveal that they are, in fact, compatible with a more general synthesis.

To enable such recombinations to succeed, even if at a superficial cursory glance they seem like they wouldn't succeed, a deep dive must be made. One must move from simple multidisciplinary thinking, in which one takes received wisdom from different disciplines and recombines ideas from them uncritically and at face value, to more complex interdisciplinary thinking, in which one dives deeply into those aspects of the disciplines necessary to decide whether a superficial judgment of "yes, the new combined idea will work" or "no, the new combined idea won't work" is justified or not. The greater the breadth of knowledge being tapped, the greater the depth of knowledge necessary to enable the tapping to be successful. This is consistent with notions regarding the simultaneous importance of divergent thinking (recombination of ideas from different disciplines) *and* convergent thinking (critically evaluating key aspects of the recombination). It is also consistent with bibliometric studies finding that papers referencing highly disciplinary knowledge (literature that historically has been cited together much more frequently than expected by chance), while at the same time citing papers that have rarely been co-cited before, are at least twice as likely to be "hits" in their field than the average paper. Novelty is prized in science but becomes especially influential when paired with familiar, conventional thought (Uzzi et al., 2013). In other words, interdisciplinary thinking combines breadth (idea recombination) and depth (reconfiguring the ideas so they will fit the new recombination), and is one reason creative research is so difficult. It requires the so-called T-shaped generalist / specialist combination researcher, who can be exceedingly difficult to cultivate (Avina et al., 2018).

Indeed, it may be in part because of the requirement of depth to enable the maximum impact of breadth that there seems to be a "sweet spot" in knowledge breadth for maximizing creativity. This echoes the common observation of a similar sweet spot in the intellectual diversity in an organization, as articulated by J. Rogers Hollingsworth, who studied outstanding research organizations in the life sciences (Hollingsworth, 2003, p. 221):

> Up to a certain point, more scientific diversity and communication increase the likelihood of breakthroughs in an organization, but when an organization has hyper-diversity with scientists focusing on so many problems, its scientific staff cannot communicate effectively to those in other fields.

Research excellence requires deep knowledge of existing paradigms *and* adaptability to new paradigms. It requires, as will be discussed in Chapter 4, the freedom, indeed the passion, to seek new paradigms and the focus gained from having mastered existing paradigms.

Value of the Next-Adjacent Possible

It is easy to dismiss the next-adjacent possible as being too *far* from the possible, as being too difficult to explore. Exploring the next-adjacent possible is indeed exceptionally risky and quickly reaches diminishing returns. As articulated by Steven Johnson (Johnson, 2011, pp. 36–38):

> The adjacent possible is as much about limits as it is about openings. At every moment in the timeline of an expanding biosphere, there are doors that cannot be unlocked yet. In human culture, we like to think of breakthrough ideas as sudden accelerations on the timeline, where a genius jumps ahead fifty years and invents something that normal minds, trapped in the present moment, couldn't possibly have come up with. But the truth is that technological (and scientific) advances rarely break out of the adjacent possible; the history of cultural progress is, almost without exception, a story of one door leading to another door, exploring the palace one room at a time. But, of course, human minds are not bound by the finite laws of molecule formation, and so every now and then an idea does occur to someone that teleports us forward a few rooms, skipping some exploratory steps in the adjacent possible. But those ideas almost always end up being short-term failures, precisely because they have skipped ahead. We have a phrase for those ideas: we call them "ahead of their time."

The next-adjacent possible is a complex space, filled with possibilities whose utility or nonutility is exceedingly difficult to distinguish—it is a space filled with dead ends. Moreover, staying within the adjacent possible is hardly a recipe for stagnation. The adjacent possible grows and becomes ever more powerful over time. Whenever something new is created, part of the formerly adjacent possible enters the possible and is therefore bounded in turn by a fresh adjacent possible. Every time a novelty occurs, the adjacent possible expands.

Nonetheless, staying within the adjacent possible is equivalent to scientists following the crowd, constrained by the same frontier of knowledge as their peers. Exploring two or more steps beyond the frontier, exploring in the next-adjacent possible, has led to many of the greatest technoscientific revolutions. Indeed, one crucial type of new knowledge that exists in the next-adjacent possible is new knowledge that is surprising and *contrarian*, as we will explore in Chapter 3 and Chapter 4. This new knowledge, if true, surprises and contradicts conventional wisdom because it is two steps away from the actual, first by recombining knowledge and then by deleting (or at least restructuring) some piece of existing knowledge or conventional wisdom in the possible that incorrectly contradicted the new knowledge.

2.3 Cycles of $\dot{Q}$ and $\dot{A}$: Special Relativity and the iPhone

This finding and matching of questions and answers, the synthetic looking upward and the analytic looking downward, takes place sometimes sequentially, sometimes simultaneously, in a complex and intricate dance. The results of the dance are almost biological, as articulated by Brian Arthur (Arthur, 2009, p. 207):

> If you examine a technology from the top down, you see it as an arrangement of connected parts interacting and intermeshing with each other to some purpose. In this sense it becomes a clockwork device—it becomes mechanistic. If you examine it mentally from the bottom up, however, from how these parts are put together, you see these as integral parts—integral organs—forming a higher, functioning, purposed whole. It becomes a functioning body—it becomes organic. Whether a technology is

> mechanistic or organic therefore depends on your point of
> view. . . . The more sophisticated and "high-tech" technologies
> become, the more they become biological.

In economic terms, the question-and-answer dance can also be thought of as positive feedback supply-and-demand cycles.

Consider first the supply-and-demand cycle in technology, viewed through the lens of the engineering method described in Chapter 1. That method begins with human-desired function that presents a demand for a new form to fulfill that function. Each such function can very likely be fulfilled by a great number of new forms—not an infinite number, of course, as new forms are constrained by nature and by the stock of existing and historically contingent forms from which to build (Rosenberg, 1974)—but a great number, nonetheless. The variety of foods grown, the kinds of farming environments, and the types of fertilizer—the possible new forms that might fulfill the function of food production are combinatorially large. But each of these forms is itself multipurpose: a tractor, for instance, can be used to prepare land for farming, but it can also be used to prepare land for construction. The availability and supply of newly engineered forms stimulates the exapting of demand for a multiplicity of new human-desired functions that those forms might *also* fulfill. The result is a powerful positive feedback cycle. Because both human desires (from which functions are drawn) and the world in which humans reside (from which forms are drawn) are seemingly infinite, human-desired function- and form-finding are also seemingly infinite.

Consider now the supply-and-demand cycle in science, viewed through the lens of the scientific method also described in Chapter 1. That method starts with facts—facts gleaned from observations. If the facts are similar, there may be one overarching explanation, but different observational circumstances may require nuance to the explanation (balls falling in air and not vacuum may need to account for air friction). These nuanced explanations are then generalized (proposed as explanations to other facts), thus leading to a proliferation of new facts that can be explained. These new facts will not be identical to the original fact, they will be different due to yet new circumstances of their observation (falling cubes or rocks of irregular shape rather than just falling balls), whose explanations will require the original explanations to be augmented in yet new ways. Because both the natural world (from which facts are drawn) and the world of ideas (from which explanations are drawn) are seemingly infinite, fact- and explanation-finding are also seemingly infinite.

To illustrate more concretely the coevolutionary dance of question- and answer-finding, we offer two examples: the theory of special relativity, drawn from science; and the iPhone, drawn from technology. Through these examples, we illustrate the give-and-take between question-finding and answer-finding in the possible and in the adjacent and next-adjacent possibles.

Special Relativity

Our first example, from science, is Einstein's 1905 theory of special relativity—the theory that simultaneously explained the principle of relativity (that physical laws are the same in all reference frames in uniform relative motion with respect to one other) and the constancy of the speed of light as observed and also as described by Maxwell's equations. The theory was the then-deepest answer (deepest explanation) to pre-existing questions which were themselves shallower answers to other questions; the theory also answered questions (predicted new facts) unrelated to the original question it intended to answer. We illustrate these relationships in Figure 2-3, organizing them around five coarsely defined and interconnected clusters of knowledge: Newtonian Mechanics, Principle of Relativity, Electromagnetism, Special Relativity, and Mass-Energy Equivalence. In every cluster, empirical facts and explanations specific to particular circumstances are depicted as being shallower toward the top of the network, while abstract explanations that are general across a range of circumstances are depicted as being deeper and toward the bottom of the network.

Newtonian Mechanics

At the top of this cluster are empirical observations (or questions), both astronomical (the motion of the stars and planets) and terrestrial (the motion of small-scale mechanical objects on earth): observations of when the sun rises and sets in the summer and winter or observations of

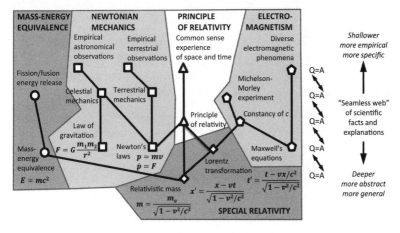

FIGURE 2-3. Special relativity and its "seamless web" of interconnecting scientific knowledge modules, with the various modules organized as questions and answers.

how fast balls roll down inclined planes. The most foundational explanations of (answers to) these empirical observations are toward the bottom: Newton's laws (momentum is the product of mass and velocity, and the time rate of change of momentum is caused by a force) and the law of gravitation (gravity exerts an attractive force between two objects that depends on their masses m_1 and m_2, the inverse square of the distance between them $1/r^2$, and the gravitational constant G). In between the two, for brevity not depicted, are numerous specific explanations in celestial and terrestrial mechanics that are themselves explained by the foundational principles below them and that explain the empirical observations above them: why planets so often have elliptical orbits or why two stones with different masses fall at the same rate.

Principle of Relativity

The foundational principle at the bottom of this cluster is Galileo's principle that the motions of bodies included in a given space are the same among themselves whether that space is at rest or moves uniformly in a straight line. This principle implies, for example, that if a spaceship is drifting along at a uniform speed, all experiments performed in the spaceship and all phenomena observed in the spaceship will appear the same as if the ship were not moving, provided, of course, that one does not look outside. This foundational principle provides an explanation (an answer) for commonsense experiences of space and time (questions), including Galileo's famous observation that a stone dropped from the mast of a sailing ship in motion falls to the base of the mast rather than to its rear. This principle, in turn, can be thought of as its own scientific fact (question) that constrained Newton's laws (as answers) to being invariant under so-called Galilean transformations between reference frames differing only by a constant relative motion.

Electromagnetism

At the top of this cluster are empirical observations of diverse electromagnetic phenomena (questions), all the way from the generation of electricity by the motion of wire coils in a magnetic field to the generation of a voltage through the charging of a capacitor. At the bottom of this cluster are Maxwell's famous equations, which explain (answer) all these observations and phenomena. Interestingly, Maxwell's equations also predicted (and thus explained) the constancy of the speed of light, c, regardless of reference frame, which was in the late nineteenth century the best explanation (answer) for the results (questions) revealed by the famous Michelson-Morley experiment.

Special Relativity

This cluster arose as a consequence of the paradox in the late nineteenth century that two central scientific facts (questions) or explanations (answers), the principle of relativity, and the constancy of c, appeared to be incompatible, even though each was consistent with a rich body of empirical observations. The constancy of the speed of light regardless of reference frame contradicted the principle of relativity as understood from Newton's laws. The two explanations that resolved this incompatibility were the Lorentz transformation, in which space and time are dilated by a factor associated with velocity relative to the speed of light, and Einstein's relativistic mass equation, in which mass is dilated by a similar factor (Feynman et al., 2011). Thus, special relativity is seen to be a foundational principle (an answer) that explains and lies below the principle of relativity and electromagnetism (as questions).

Mass-Energy Equivalence

This cluster is perhaps that which special relativity is most famous for explaining. From relativistic mass (an answer to the two clusters), Einstein deduced the new fact (or question) of mass-energy equivalence. That new fact was itself an explanation (or answer) to the question of energy release on nuclear fission and fusion events and ultimately of energy release from nuclear weapons.

Overarching Observations

Before we leave the theory of special relativity, we make three overarching observations about the fluidity, spontaneity, and boundary-crossing nature of the co-evolutionary dance of question-and-answer finding in the story of the theory of special relativity.

First, the hierarchy of fact and explanation just discussed does not necessarily follow from the time ordering in which the facts and explanations were initially discovered. On the one hand, the fact of the constancy of c preceded its explanation in special relativity: the constancy of c was an existing "question" whose scientific answer was found afterwards. On the other hand, the principle of mass-energy equivalence preceded the observation of the energy release in fission / fusion but explained it when it was observed: mass-energy equivalence was an answer to one question, but then was generalized to answer another question to predict observations completely different from the observations it was originally intended to answer.

Second, as will be discussed later in Chapter 3, though these knowledge modules reside in the scientific knowledge network, their creation and extension made heavy use of (and led to new), knowledge modules in the technological knowledge network. The telescope was necessary for the astronomical observations that led to the facts that celestial mechanics sought to explain. The global positioning system on which modern location tracking is based depends on an understanding of special relativity. Nuclear power plants make use of the principle of mass-energy equivalence.

Third and finally, while some of the new questions and answers associated with special relativity can be thought of as having been found in the adjacent possible, many cannot. Newton's laws required him to invent a new calculational technique (calculus) from the next-adjacent possible *en route* to explaining the celestial and terrestrial observations he set out to explain. Similarly, the Michelson-Morley experiment required overcoming significant instrumentation challenges before the speed of light could be determined with sufficient accuracy for the fact of its constancy to be established.

The iPhone

Our second example, from technology, is the iPhone during the years 2004–2007 when it was taking shape as a "gleam in Steve Jobs' eye" (Merchant, 2017). Like special relativity, the iPhone was at the center of a network of questions and answers—though here we use the language of problems and solutions. It played both the role of a form that solved particular functional problems as well as the role of a function that was a problem waiting for forms to be found that would solve that problem. We organize this example into the technological knowledge problem-and-solution network illustrated in Figure 2-4, with the following notations.

If a knowledge module (serving as part of a solution) has an arrow connecting up to a knowledge module (serving as a problem) above it, the solution is motivating the finding of a problem. The leftmost arrow

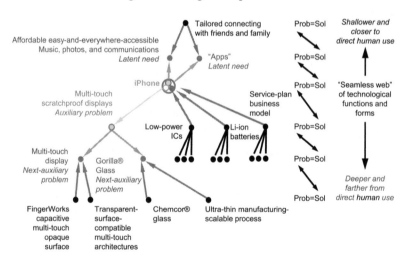

FIGURE 2-4. The iPhone and its "seamless web" of interconnecting technological knowledge modules, 2004–2007, with the various modules organized as problems and solutions.

pointing upward from the iPhone reflects the fact that the iPhone was a solution that Steve Jobs hoped would solve some problem, and he was visionary enough to see that it might be the latent problem of "affordable, easy-and-everywhere-accessible music, photos and communications."

If a knowledge module (serving as a problem) has an arrow connecting down to knowledge modules (serving as a solution) below it, the problem is motivating the finding of a solution. The leftmost arrow pointing downward from the iPhone reflects the fact that the "gleam in Jobs' eye" iPhone was a problem that he hoped a multitouch scratch-proof display could solve.

Of course, if an arrow connects a knowledge module downward to another knowledge module, it is simultaneously connecting that other knowledge module upward. Problems and solutions are connected as pairs, and which is finding which can sometimes be a matter of perspective: they are "finding each other." The gesture-recognition company FingerWorks was about to go out of business for lack of problems that its capacitive multitouch surface solution could solve, so it was on the prowl for problems, while Apple was on the prowl for a solution to the problem of multitouch displays. They were looking for each other. Corning, the glass and ceramics company, had not been very successful at finding customers for its Chemcor ultrahard glass (fifteen times harder than regular glass), while Apple was on the prowl for a solution to the problem of a scratchproof multitouch display. Again, they were looking for each other. From the perspective of FingerWorks and Corning, their "solutions looking for problems" were exapted by the new problems of multitouch display and Gorilla Glass® that Apple presented to them. From the perspective of Apple, its "problems looking for solutions" were resolved by the technologies that FingerWorks and Corning presented to it.

Though the end result is always that a problem and solution have found each other, the direction of one's perspective is important and implies different types of innovation risk. An upward-facing perspective means solutions hunting for problems (needs or markets), so there is "market risk" associated with whether the market need you think you have found is a real market need or not. A downward-facing perspective means problems hunting for solutions, so there is "technical risk" associated with whether the solution you think you have found really solves your problem or not. If one has simultaneously both an upward- and downward-pointing perspective, then one is faced with both solutions hunting for problems and problems hunting for solutions, so there is both market and technical risk. It is a testament to Steve Jobs' courage that he took on both kinds of risk at multiple levels of the iPhone knowledge hierarchy.

Moreover, the degree of risk increases with the degree to which one is looking for problems or solutions in the possible or in the adjacent and next-adjacent possibles. The iPhone itself, at the center of Figure 2-3, was in the next-adjacent possible—requiring many auxiliary problems and solutions in the possible and adjacent possible to be found. The universe of forms that ultimately solved the problem of the iPhone, and the universe of functions to which the iPhone ultimately provided solutions—these all surround the iPhone; the farther away, the closer to the possible they are. In what follows, we give two views of the creation of the iPhone: one view begins with the possible and works its way "inward" to the adjacent and next-adjacent possible, ending at the iPhone itself; another view begins with the iPhone, in the next-adjacent possible, and works its way in the other direction, "outward" to the adjacent possible and the possible.

From the Possible to the Adjacent and Next-Adjacent Possibles

Let us start by working inward from the "possible" to the "adjacent possible" from the already-existing knowledge modules at the top and bottom, inward to not-yet-existing knowledge modules closer to the center.

At the bottom far left are two knowledge modules that already existed at the time: a capacitive multitouch opaque surface (suitable for trackpads) pioneered by FingerWorks (purchased by Apple in 2005) and architectures that were compatible with multitouch on transparent surfaces and hence suitable for displays. Whether these two already-existing knowledge modules could actually be combined into a multitouch display was not known in 2005; but, in hindsight, we now know that they could be, so they both connect upward into a multitouch display knowledge module that is in the adjacent possible.

At the bottom not-so-far left are two additional knowledge modules that also already existed at the time: a Chemcor ultrahard-but-thick glass that had been developed by Corning decades earlier and processes that one might use for making this glass ultrathin, as would be necessary for a handheld display. Again, whether these two already-existing knowledge modules could actually be combined into what we now know as Gorilla Glass® was not known in 2006; but, in hindsight, we now know that they could be, so they both connect upward into a new Gorilla Glass® knowledge module that is in the adjacent possible.

Importantly, just because these two composite knowledge modules, the multitouch display and Gorilla Glass®, are in the adjacent possible doesn't mean they would actually ever be realized. It could have been that the problem that they solve was never recognized or deemed important, so there would be no motivation to bring those knowledge modules from the adjacent possible into the possible. In an alternative history, Steve Jobs might not have committed to making an iPhone. He might never have bought FingerWorks nor pushed them to realize a multitouch display, and he might never have pushed Corning to develop Gorilla Glass®. Both multitouch displays and Gorilla Glass® could have ended up being "latent solutions" that never found problems to solve and hence forever remained in the shadowy and unrealized adjacent possible.

At the far right, we also draw a number of other already-existing knowledge modules that eventually made it into the iPhone: low-power integrated circuits, lithium-ion batteries, and the service plan business model. There are many more, of course, and we draw these only to illustrate that not all the knowledge modules that the iPhone was trying to draw on in 2006 were in the adjacent possible—many already existed and were in the "possible."

At the top is a black dot signifying our basic human desire to "connect with friends and family." There were lots of solutions to this "problem" in 2006 before the iPhone (including, of course, just visiting and talking face-to-face), so there could be a slew of black dots that connect upward to this black dot that we have not drawn so as not to clutter the diagram. There is also a slew of dots that could connect upward to this black dot because there are lots of unrealized but imaginable ways of solving the problem of connecting to friends and family. One such dot is the one that is labeled "affordable, easy-and-everywhere-accessible music, photos, and communications." This dot is a "latent need": a need that most people in 2006 did not realize they had, but which subsequent history has shown they *did* have—otherwise Apple would not have grown to have become among the world's most valuable companies. Not only was this need latent, it was, we now know, realizable. Not all latent needs are, however—many are unrealizable with current technologies—that is, there might not be any already-existing knowledge modules that together could create a solution to meet the latent need. Dorothy might want to get home to Kansas, but that latent need might or might not be realizable. Realizable or unrealizable, however, such latent needs do belong in the adjacent possible. The reason is the same, just in reverse, as the reason we include latent solutions in the adjacent possible: a latent need with no solution does not make it any less in the adjacent possible than a latent solution with no problem it solves.

From the Next-Adjacent and Adjacent Possibles to the Possible

Now, working inward out, from the "next-adjacent and adjacent possibles" to the "possible," let us start with the open circle labeled "iPhone."

From the iPhone circle working upward, we see the knowledge modules that are the latent needs or latent problems that the iPhone would solve if it could be realized: "affordable, easy-and-everywhere-accessible music, photos, and communications" and "apps." As discussed above, these latent needs are in the adjacent possible. Then, working upward one more level, we see the actual need that would be met if these latent needs could be realized: "tailored connecting with friends and family."

From the iPhone circle working downward, we see knowledge modules that would need to be realized to create an iPhone. Again, we have not been exhaustive so as to keep the diagram from being too cluttered. On the right are the three knowledge modules mentioned above (low-power integrated circuits, lithium-ion batteries, and a service plan business model) that already existed in 2006. But diagonally below and to the left is a knowledge module (multitouch scratchproof displays) that did not exist in 2006. From this perspective, it represents what the Hungarian mathematician George Polya called an "auxiliary problem": secondary problem that, if it could be solved, would enable the solution of the primary problem. Indeed, not only is this an auxiliary problem; it is a nontrivial-to-solve auxiliary problem—in fact, it was insoluble using knowledge modules that existed in 2006. One needed the two other knowledge modules (the multitouch display and Gorilla Glass®) that were discussed above. These two knowledge modules might be thought of as next-auxiliary problems in the sense that they themselves are auxiliary problems for the multitouch scratchproof display, which, in turn, is an auxiliary problem for the iPhone. And now, finally, working downward comes to an end because these two knowledge modules are in the adjacent possible—they themselves were realizable using knowledge modules that existed in 2006.

Overarching Thoughts

Before we leave the iPhone as an example, we make two overarching observations about the fluidity, spontaneity, and boundary-crossing nature of the co-evolutionary dance of question-and-answer finding in the creation of the iPhone.

Just as with the example of special relativity from science, in this example of the iPhone, the hierarchical ordering of functions and forms that fulfill those functions does not necessarily follow from the time ordering in which the functions and forms were initially discovered. On the one hand, the desire to create an iPhone preceded, and provided the impetus to develop, the multitouch scratchproof display. The iPhone was a "problem" needing a "solution." On the other hand, low-power integrated circuits and lithium-ion batteries preceded, and helped enable, the iPhone. Low-power integrated circuits and lithium-ion batteries had already solved other problems, including other mobile electronics like simple (non-smart) phones, music players, digital cameras, and laptop computers. Having solved those other problems, they could then be exapted to address new problems.

As with the example of special relativity, in this example, the creation of the iPhone made heavy use of both scientific and technological knowledge. A scientific understanding of surface-stress hardening helped enable the Chemcor scratchproof glass technology. And the iPhone as a technology is in turn enabling creation of new science. As one example, through the iPhone's embedded sensors and location tracking and through its connectivity with social media, the measuring and finding of new "social" facts may lead to breakthroughs in our scientific understanding of human social behavior.

2.4 Recapitulation

To recapitulate, in this chapter we outlined our second stylized fact associated with the nature of research: that human technoscientific knowledge is organized into loose hierarchically modular networks of question-and-answer pairs and that these questions and answers evolve in an intricate dance to create new question-and-answer pairs.

Question-finding and answer-finding can be mapped to the six mechanisms of the technoscientific method discussed in Chapter 1. In the engineering method half of the technoscientific method: function-finding is the finding of new questions, new human-desired functions; form-finding is the finding of new answers, new forms that fulfill those human-desired functions, and exapting is the finding of new questions, new human-desired functions that are fulfilled by existing forms. In the scientific method half of the technoscientific method: fact-finding is the finding of new questions, new facts of human interest to explain; explanation-finding is the finding of new answers, new explanations that explain those facts; generalizing is the finding of new questions, new facts predicted by explanations of other facts. Importantly, answers of a scientific nature can lead to questions of an engineering nature. Proposed scientific explanations of previously observed scientific facts generalize and predict potential new scientific facts. Observing these potential new scientific facts is an engineering function, often requiring the finding of new technological forms tailored to enable the new observations. Just as importantly, answers of an engineering nature can lead to questions of a scientific nature. A technological form that fulfills a human-desired function, as it performs that function, can reveal unusual phenomena that become scientific facts ripe for explanation—scientific questions ripe for scientific answers.

New questions or answers are not equally easy to find. They are harder to find the further they are from existing answers and questions—from what might be called the "possible." Nearest are question-and-answer pairs that exist and are matched to each other in the "possible" but with room for optimization and improved matching. Further are *latent* question and answers in the "adjacent possible": where a knowledge-recombination step is required. Further still are questions and answers in the "*next*-adjacent possible": one combinatorial step removed from the "adjacent possible" and *two* combinatorial steps removed from the "possible." Moreover, the further the new questions and answers are from the possible, not only is the connection more difficult to make but also the greater the unexpectedness of the connection, the greater the possible surprise, and the greater the potential for overturning conventional wisdom—what we will identify in Chapter 3 with "paradigm creation."

Just as the various mechanisms of the technoscientific method are cyclical, the various mechanisms of question-and-answer finding are cyclical. When seeking an answer to one question, an answer to another question often emerges: X-rays were the answer to the question of what caused the fluorescence on a screen too far away to be affected by cathode rays, but they ultimately answered new questions of how to image the inner organs of the human body. Or, when seeking a question that existing answers might help answer, necessary auxiliary answers are catalyzed: optical communications was a new question that lasers might be able to answer, but only if one could find a low-loss medium, such as ultrapure glass fibers, by which the laser light could controllably propagate. Thus, neither question- nor answer-finding is more important or "leads" the other; they fuel each other. As we will discuss later in Chapter 4, effective nurturing of research means nurturing whichever is ripest for advance—sometimes it is question-finding, sometimes it is answer-finding, and sometimes it is both nearly simultaneously.

Special relativity and the iPhone are two iconic examples of both the interplay between finding questions and answers as well as their being found in the possible, adjacent possible, and next-adjacent possible. In special relativity, the constancy of c was a question answered by the theory of special relativity; the theory of special relativity, in turn, was generalized to become an answer to a completely different question regarding the energy release in fission / fusion. The Michelson-Morley experiment was an answer to the question of whether the speed of light is constant or not, but it was in the next-adjacent possible because it first required overcoming significant instrumentation challenges (auxiliary problems) in the adjacent possible. The iPhone was a question that depended on low-power integrated circuits and lithium-ion batteries as answers that were originally developed to answer other questions, such as how to make laptop computers and other mobile computational devices. The multitouch scratchproof display was another piece of the answer to the question of how to make an iPhone, but it was in the next-adjacent possible because it first required overcoming significant challenges (auxiliary problems) in the adjacent possible associated with ultrahard Gorilla Glass®. The interplay between question-finding and answer-finding in the possible and in the adjacent and next-adjacent possibles is truly an intricate and beautiful dance.

3

The Punctuated Equilibria of Knowledge Evolution

Surprise and Consolidation

In the previous two chapters, we described two perspectives on technoscientific knowledge and its evolution. One perspective centered on science and technology as distinct and coequal types of knowledge, evolving via the various mechanisms of what we call the technoscientific method. The other perspective centered on questions and answers nested in a seamless web of knowledge, evolving via question-and-answer finding. In this chapter, we offer a third perspective, one that cuts across the first two: that the evolution of technoscientific knowledge, whether from the perspective of the technoscientific method or of question-and-answer finding, marches to a drummer whose rhythm is not constant but highly variable.

This third perspective is our third stylized fact: that technoscientific knowledge evolves via punctuated equilibria in which periods of relatively more gradual advance are punctuated by relatively more sudden advances. Here, we borrow language from evolutionary biology (Gould & Eldredge, 1993), in which the gradual and more continuous adaptation of populations to their environmental niches is occasionally punctuated by more discontinuous speciation events—often the result of a subpopulation becoming separated from its larger population and thrust into a new environment. A classic example of such a discontinuous event in evolutionary biology is when one branch of dinosaurs suddenly developed the capability for flight that ultimately led to the modern bird—an example we used in introducing the term "exapting" in Chapter 1. Applied to knowledge evolution, as illustrated highly schematically in Figure 3-1, we think of more continuous adaptation as the strengthening and "consolidation" of conventional wisdom, and of more discontinuous punctuation events as "surprise," an unexpected challenge to (and overturning of) conventional wisdom. These are also the two kinds of learning, discussed in Chapter 1, for which human culture selects when deciding what kinds of $\dot{S}$ and $\dot{T}$ to pursue: learning by consolidation and learning by surprise.

Surprise and consolidation go hand in hand, and are both important to the advance of knowledge. As discussed in the introduction,

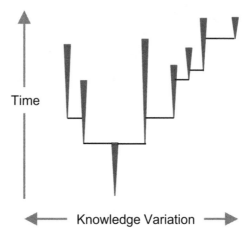

FIGURE 3-1. Schematic depiction of punctuated equilibria in knowledge evolution.
Credit: Adapted and modified from "The model of punctuated equilibrium contrasted against the phyletic gradualism model of speciation," Miguel Chavez / Wikimedia Commons / CC BY-SA 4.0.

however, an opposing belief is widespread but mistaken: the belief that the goal of research is to strengthen and consolidate conventional wisdom, rather than to surprise and possibly overturn it. In this chapter, we correct that belief. We describe, instead, the different but essential roles that *surprise* and *consolidation* play in advancing knowledge: how the first can be associated with the creation, and the second, the extension, of what we will call sociocultural technoscientific paradigms; and how they play off of each other in powerful cycles in which each spawns the other.

We start by introducing paradigms as the mediators of punctuated equilibria. Paradigms are combinations of knowledge "put to work" to advance knowledge—a kind of metaknowledge of how to *use* knowledge modules to accomplish dynamic *changes* to knowledge modules. The creation of a new paradigm represents a relatively sudden and surprising break from conventional wisdom, while the extension of an existing paradigm represents a relatively more gradual consolidation and strengthening of conventional wisdom. Then we discuss the interplay between surprise and consolidation—between paradigm creation and extension. The creation of a new paradigm creates "open space" for the extension of the paradigm—for it to strengthen, and be strengthened by, the knowledge on which it draws. The extension of a paradigm, in turn, sows the seeds for the creation of yet newer paradigms. Finally, we discuss examples of cyclical surprise and consolidation, or cyclical paradigm creation and extension, from the history of artificial lighting.

3.1 Paradigms as Mediators of Surprise and Consolidation

What do we mean by paradigms as combinations of knowledge "put to work" to advance technoscientific knowledge? To describe what we mean, we distinguish, as discussed in Chapter 1, between the static repositories of technoscientific knowledge (S and T) and the use of those repositories to bring about dynamic change in technoscientific knowledge ($\dot{S}$ and $\dot{T}$). As static repositories, scientific and technological knowledge are qualitatively different and separate from each other. Each is organized into its own seamless web of question-and-answer pairs. As dynamic repositories, however, scientific and technological knowledge are not independent; indeed, they are highly interdependent. For example, the scientific fact of the constancy of the speed of light is independent of technology, but its *establishment* as a scientific fact depended on the sophisticated technology embodied in the Michelson-Morley experiment mentioned in Chapter 2. The technological form of the ultra-hard Gorilla Glass® used in the iPhone is independent of science, but its *establishment* as a technological form depended on the sophisticated science of surface hardening by ion exchange.

The "doing" of $\dot{S}$ and $\dot{T}$ thus depends on knowledge scattered across science and technology. Moreover, as discussed in Chapter 1, it also depends on culture, which provides the selection pressure for what $\dot{S}$ and $\dot{T}$ to pursue. Thus, the doing of $\dot{S}$ and $\dot{T}$ depends on knowledge scattered across science, technology, *and* culture. And the doing of a particular mix of $\dot{S}$ and $\dot{T}$ depends on a particular mix or holistic combination of science, technology, and culture. In this book, we call these mixes "paradigms." Paradigms are the combined sociocultural and technoscientific knowledge at play as new questions and answers are deemed worthy of finding, as the choices of what S and T to make use of are made, and as the arduous processes of $\dot{S}$ and $\dot{T}$ are actually "done." Paradigms are combined sociocultural *and* technoscientific constructions, binding a social community of technoscientists together in common agreement on how to move technoscientific knowledge forward.

How do paradigms move technoscientific knowledge forward? They do so by mediating the mechanisms of science-and-technology coevolution discussed in Chapter 1 and the question-and-answer coevolution discussed in Chapter 2. At a low level, paradigms mediate what particular technoscientific knowledge modules to draw on to advance knowledge. Are telescopes the instrument of choice to advance a particular knowledge domain? What sorts of improvements in telescopes would be most helpful to advance that particular knowledge domain? What scientific theories and engineering tools might be brought to bear to make those improvements? At a high level, paradigms mediate which mechanism of the technoscientific method to focus on at any given time. Is there more opportunity in finding new functions, in finding forms to fulfill functions, or in exapting existing forms to new functions? Is there more opportunity in finding new facts, in finding new explanations for existing facts, or in generalizing existing explanations to other facts?

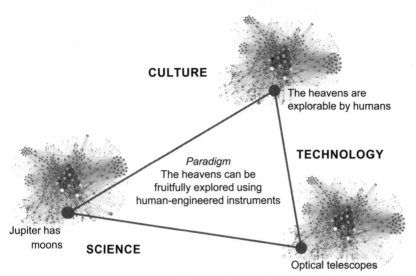

CULTURE

The heavens are explorable by humans

TECHNOLOGY

Paradigm
The heavens can be fruitfully explored using human-engineered instruments

Jupiter has moons

SCIENCE

Optical telescopes

FIGURE 3-2. An example of a sociocultural technoscientific paradigm spawned by Galileo's discovery of the moons of Jupiter. *Credit:* Network graphics Martin Grandjean / Wikimedia Commons / CC BY-SA 3.0.

In the example illustrated in Figure 3-2, the new paradigm spawned by Galileo's discovery of the moons of Jupiter in 1610 might be paraphrased as "the heavens can be fruitfully explored using human-engineered instruments." Oversimplifying for illustrative purposes, we might say the paradigm draws on and connects three knowledge modules, each residing in either the science, technology, or culture knowledge networks: the emerging technology of optical telescopes that originated in the Netherlands, the new scientific fact found by Galileo that Jupiter has moons, and the new cultural value that the heavens are explorable by humans. In the previous paradigm, it would have made no sense to train a telescope on the heavens given Aristotle's pronouncement that celestial objects were perfect and immutable spheres. In the new paradigm, opportunity has opened up for the exploration of other objects in the heavens and for the customization and improvement of telescopes specifically for astronomical exploration.

This example also makes vivid the role of paradigms in the punctuated equilibria of knowledge evolution. The punctuation events are the creation of new paradigms—relatively sudden and surprising breaks from conventional wisdom. The periods of equilibrium in between are the extension of those paradigms—the gradual strengthening and consolidation of the new paradigm as it itself becomes conventional wisdom. The punctuation events are like exclamation marks, or surprises, between periods of equilibrium as the sociocultural technoscientific community grows accustomed to the surprise and consolidates it into a new conventional wisdom.

Paradigms as Exemplars

How do paradigms mediate the gradual strengthening and consolidation of knowledge? One powerful way is by providing an exemplar for how a particularly difficult problem was once solved and, thus, how similar problems might be solved. Often by seeing and deeply understanding an exemplar instance of the finding of a new question or answer, one can generalize to the finding of similar questions or answers. Past practice, beginning with the breakthrough exemplar, is a powerful teaching tool.

This way of thinking about paradigms also highlights another difference between the static structure (S and T) and the dynamical evolution ($\dot{S}$ and $\dot{T}$) of science and technology.

The static structure of science and technology is largely formal and codified into knowledge modules, as we have seen. Scientific explanations are most elegant when they are codified into mathematical equations such as $E = mc^2$ or $F = ma$. Technological forms are equally codified, instantiated as they are into concrete artifacts and the processes by which those artifacts are made and used. It is the job of textbooks to elucidate the static structure of S: How does the theory of special relativity explain the equivalence between energy and mass? It is the job of the instantiated technological form to work and fulfill its function: the telescope fulfilling its function of magnifying distant objects.

In contrast, the dynamic evolution of science and technology is less formal and less codified. Knowing *how* to use the static structure of S and T to artfully do $\dot{S}$ and $\dot{T}$ requires a kind of tacit knowledge. And, like most (if not all) tacit knowledge, it can often only be learned by example. As articulated by Thomas Kuhn (Kuhn, 1974, p. 9):

> Students of physics regularly report that they have read through
> a chapter of their text, understood it perfectly, but nonetheless

had difficulty solving the problems at the end of the chapter. Almost invariably their difficulty is in setting up the appropriate equations, in relating the words and examples given in the text to the particular problems they are asked to solve. Ordinarily, also, those difficulties dissolve in the same way. The student discovers a way to see his problem as like a problem he has already encountered. Once that likeness or analogy has been seen, only manipulative difficulties remain.

In other words, the "seeing" of a paradigm is not trivial. It is similar to the current challenge in machine / deep learning of extracting from neural net classification algorithms, once trained, *how* they classified and then to generalize to new classifications. Seeing a paradigm, really understanding it, is akin to the "deep craft" discussed by Brian Arthur (Arthur, 2009, p. 159):

> Real advanced technology—on-the-edge sophisticated technology—issues not from knowledge but from something I will call deep craft. Deep craft is more than knowledge. It is a set of knowings. Knowing what is likely to work and what not to work. Knowing what methods to use, what principles are likely to succeed, what parameter values to use in a given technique. Knowing whom to talk to down the corridor to get things working, how to fix things that go wrong, what to ignore, what theories to look to. This sort of craft-knowing takes science for granted and mere knowledge for granted.

Paradigms thus provide exemplars that, once seen, become a powerful kind of knowledge in and of themselves—knowledge, in large part tacit, that patterns the finding of new and similar questions and answers and the mapping of the solution of one problem to the solution of another. Because of their tacit nature, paradigms are much less easy to crisply define and measure than the knowledge modules they connect and use. However, not all things that are important are yet definable and measurable. Even from our perspective as physical scientists and engineers used to great knowledge codification and mathematical formality, we recognize the importance of the sociocultural and tacit aspects of human knowledge.

Paradigms as Scaffolding

Another way to think about paradigms is as a kind of metaknowledge, knowledge about knowledge. Paradigms represent metaknowledge that draws on but overarches the static technoscientific networks of existing knowledge modules. Paradigms represent metaknowledge of how to *use* knowledge modules to accomplish dynamic *changes* to human knowledge. Knowledge modules are like nouns—particular chunks of technoscientific knowledge, both concrete and abstract. Paradigms, in contrast, are like sentences—the use of knowledge modules to create or extend knowledge modules.

One can also think of knowledge modules and the paradigms used to create or extend them as a building and the scaffold used to build it. The scaffold is used to build the building, but once built, the building can be used even if the scaffold is disposed of. Likewise, a paradigm, once created, is used to create knowledge modules but, once created, knowledge modules can be exploited "as is" even if the paradigm goes "dormant." In the case of scientific knowledge, the fact that Jupiter has moons was created by a paradigm that included, for example, use of the optical telescope, a module in the technological knowledge network. But, once that fact was established, the telescope is no longer needed to support the paradigm—the fact of Jupiter's moons stands by itself in the scientific knowledge network. In the case of technological knowledge, the Chemcor ultrahard glass on which Gorilla Glass® and the iPhone were ultimately based is a module in the technological knowledge network. The creating of that Chemcor ultrahard glass required a scientific understanding of surface-stress hardening by size-mismatched ions. But once Chemcor ultrahard glass itself has been produced, it can be used "as is" in the iPhone—the scientific understanding of *why* it works is no longer needed.

Of course, if one were not content with knowledge as is but wanted to continue to create or extend knowledge—to explore and not just exploit—one would need to re-erect the scaffold. If one wanted to reverify the fact of the moons of Jupiter to a skeptic, or extend that fact to other similar facts (that other planets have moons), one would need to reactivate the paradigm that includes use of the optical telescope. If one wanted to extend Gorilla Glass® to higher levels of hardness, one would need to reactivate the paradigm that includes the scientific understanding of surface-stress hardening. Indeed, paradigms do not commonly go dormant. Knowledge modules are like living things, always evolving and growing, and paradigms are the vehicles for that evolution and growth. Moreover, the paradigms themselves, because they draw on these constantly evolving and growing knowledge modules, they themselves are constantly evolving and growing. Paradigms, which coevolve *with* underlying knowledge modules, thus play a dual role: as scaffolding that acts to create new knowledge and as beneficiaries of the creation of that new knowledge.

Moreover, if a paradigm *has* gone dormant, it can be difficult to resurrect. Resurrecting a paradigm is not simply a matter of reading textbooks. Textbooks reveal knowledge modules only as they exist—final, elegant results crafted into a logical and coherent structure—not the tortuous paths that might have been required to create the knowledge module to begin with. Resurrecting a paradigm involves resurrecting the metaknowledge associated with those tortuous paths, paths that give insights into how a knowledge module might or might not be extended or created. It is a very different thing to know something versus to know how that something was brought about.

Surprise and Consolidation as Disconfirming or Confirming Belief or Disbelief

What exactly do we mean by "relatively sudden and surprising breaks from conventional wisdom," or by "relatively more gradual consolidation and strengthening of conventional wisdom?" We mean that conventional wisdom, including both its beliefs and disbeliefs, has either been disconfirmed or confirmed. To see this, imagine we have found a potential new scientific fact or explanation, or a potential new technological function or form. By potential, we mean that they haven't yet been subjected to rigorous test in the real world. The scientific fact might be the result of a mistaken measurement, and the scientific explanation could have been verified by a mistaken calculation. The technological function might be a mistaken guess at what humans desire, and the technological form might just be a design not yet built and verified.

Whatever the potential new knowledge is, imagine that we subject it to two evaluations. The first, "prior" evaluation is when the knowledge is still potential, *before* it has been subjected to rigorous testing in the real world. Is it likely, according to conventional wisdom, that the potential knowledge will have utility? By utility, we use the broad definition introduced in Chapter 1—that the potential knowledge "works" and has an impact on human desires. The second, "posterior," evaluation is when what we have done is no longer potential but real—*after* it has been subjected to rigorous test in the real world. Does the knowledge have actual real-world utility?

The prior evaluation of utility we call u_{prior}—our best guess, based on prior conventional wisdom, of the utility of the potential knowledge. The posterior evaluation of utility we call u_{post}, after it has been subject to rigorous testing in the real world. If u_{prior} and u_{post} are the same, then prior conventional wisdom was correct and does not need to be updated. Conventional wisdom has been *consolidated*, and the paradigm on which the new knowledge was based has been extended and strengthened. But if u_{prior} and u_{post} are not the same, then prior conventional wisdom was incorrect and *does* need to be updated. Conventional wisdom has been *surprised*, and the paradigm on which the potential new knowledge was based must be revised and, in an extreme case, overturned and replaced by another paradigm.

For concreteness and simplicity, consider that the prior guessed and posterior actual utilities can be either high or low. Then there are four possible combinations, as in the four-square illustration of Figure 3-3.

- **Confirm Disbelief.** Conventional wisdom "disbelieves" that the potential knowledge will be useful, and, after further playing out, the knowledge is found indeed not to be useful. For example, preliminary observations that nuclear fusion could take place at room temperature (so-called "cold fusion") were disbelieved by conventional wisdom and were ultimately found to be incorrect (Close, 2014). Conventional wisdom's disbelief was confirmed, and the paradigm associated with that conventional wisdom was consolidated and strengthened. The scientists who made and believed the preliminary observations made a risky, and ultimately failed, attempt to overturn conventional wisdom and thus create a new paradigm.
- **Confirm Belief.** Conventional wisdom "believes" that the potential knowledge will be useful, and, after further playing out, the knowledge is indeed found to be useful. This, again, is a consolidation of

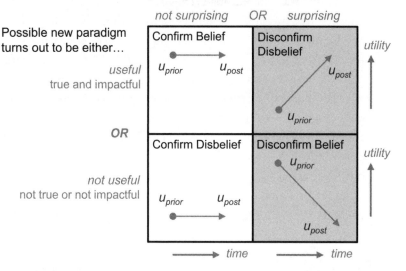

The *utility* of the possible new paradigm turns out to be either...

not surprising   OR   *surprising*

Possible new paradigm turns out to be either...

	not surprising	surprising
useful true and impactful	Confirm Belief $u_{prior} \longrightarrow u_{post}$	Disconfirm Disbelief u_{post} ... u_{prior} utility
OR		
not useful not true or not impactful	Confirm Disbelief $u_{prior} \longrightarrow u_{post}$	Disconfirm Belief u_{prior} ... u_{post} utility

$\longrightarrow$ time $\longrightarrow$ time

FIGURE 3-3. Fourfold typology of the utility of potential knowledge and whether that utility is surprising or not.

conventional wisdom and a consolidation and strengthening of the paradigm associated with that conventional wisdom. In some sense, this is also low-hanging fruit whose initial likelihood of utility is high (because it is consistent with conventional wisdom), and whose final utility is indeed high. This does not by any means imply that the knowledge was easy to come by—it could and very often will have involved difficult puzzles, manipulations, and deep and hard-won consolidation of conventional wisdom and the current paradigm. A new quantum computer may be completely consistent with quantum mechanics, but that doesn't make it easy to understand or build.

- **Disconfirm Belief.** Conventional wisdom "believes" that the potential new knowledge will be useful, but, after further playing out, the knowledge is found not to be useful. An example is the Ehrenfest or Rayleigh–Jeans "ultraviolet catastrophe": the prediction that an ideal black body at thermal equilibrium should emit more energy at higher frequencies, a prediction that was completely plausible based on the known classical physics of the late nineteenth century and early twentieth century. That prediction, however, did not pass the rigorous test of conservation of energy. Thus it didn't "work" and was disconfirmed. The problem with the idea was only later resolved by Planck via quantization, ultimately leading to the development of the completely new paradigm of quantum mechanics. This was a surprise—a failed application of the current paradigm, indicative of its limits. But, with a properly prepared mind, it can also spur the creation of a new paradigm that creates new wisdom different from prior conventional wisdom.

- **Disconfirm Disbelief.** Conventional wisdom "disbelieves" that the potential new knowledge will be useful, but after further playing out, the knowledge is indeed found to be useful. This also represents surprise and, because of the utility of the new knowledge, is likely to directly result in the creation of a new paradigm and the destruction of an old paradigm. This is also classic creativity, whereby revolutionary ideas that run counter to conventional wisdom are ultimately proved useful (Simonton, 2018). Flying machines heavier than air, evolution by natural selection, quantum mechanical action at a distance, wave-particle duality, the theory of continental drift, efficient blue LEDs fabricated from highly defective semiconductors—all these ideas were initially disbelieved but later proved correct and useful, and, hence, changed the way we think and do.

The two "confirm" outcomes are the "consolidate" outcomes—outcomes that consolidate and extend current paradigms and conventional wisdom. The two "disconfirm" outcomes are the "surprise" outcomes—outcomes that surprise the current paradigm and overturn the conventional wisdom. Previous conventional wisdom thought to be true must be deleted or unlearned—reminiscent of the saying sometimes attributed to Mark Twain:

> What gets us into trouble isn't what we don't know; it's what we know for sure that just ain't so.

Both surprise outcomes are shaded in gray in Figure 3-3 because we associate both with what we will call in Chapter 4 the "metagoal" of research—the seeking of surprise. Surprise is synonymous with disconfirmation—either disconfirmation of belief *or* disbelief that the potential new knowledge has utility. However, we call special attention to the disconfirm-disbelief outcome that combines surprise *and* utility. This outcome might also be called "useful learning" or "implausible utility" because the ultimate high utility was initially implausible to conventional wisdom (Tsao et al., 2019). And this outcome is also the one often associated with classic creativity: utility without surprise is insufficient, because utility consistent with conventional wisdom does not open up new possibilities the way utility with surprise does; but surprise without utility is also insufficient, because surprise that something *isn't* useful does not necessarily reveal what *is* useful. Nonetheless, all surprise represents significant learning. When new knowledge is surprising, it is a harbinger of new territory for exploration and of new ways of doing and thinking.

Surprise and Consolidation throughout the Technoscientific Method

Importantly, none of the mechanisms of the technoscientific method (from fact-finding to exapting) have monopolies on either paradigm creation or extension. Surprise and consolidation occur throughout the individual mechanisms of the scientific and engineering methods.

- **Surprise in the scientific method.** In fact-finding, the unexpected and almost unbelievable new fact initiates major scientific advance. Think of the discovery of Jupiter's moons, of an atom's mass being concentrated in a tiny nucleus, of superconductivity, of the fractional quantum Hall effect, and of dark matter and energy. In explanation-finding, the unexpected and almost unbelievable new theory sweeps away dusty old ways of thinking: the discovery of Newton's laws, of special and general relativity, of Cooper pairing in superconducting metals, and of quantum mechanics. Finally, in generalizing, the unexpected and almost unbelievable new generalization opens new opportunities, for example, the generalization of special relativity to energy release on radioactive decay, of the quantum mechanics of atoms to molecules and solids, and of the thermodynamics of steam engines to all energy conversion processes.
- **Surprise in the engineering method.** In function-finding, it is the unexpected and almost unbelievable new human-desired function

that initiates major engineering advance: the discovery of the vast human need for computation functionality that far exceeded Watson's estimate of "a world market for maybe five computers." In form-finding, the unexpected and almost unbelievable new form sweeps away old forms: the transistor sweeping away the vacuum tube, the automobile sweeping away the horse and buggy, and the airplane sweeping away passenger ships. In exapting, it is the unexpected new use that opens new opportunities: the exaptation of microwave radars to cooking, of nuclear magnetic resonance technology to medical imaging, and of lasers to optical fiber communications.

- **Consolidation in the scientific method.** In fact-finding, consolidation happens when the new fact is analogous to and extends previous facts, such as moons of other planets after the moons of Jupiter have been found. In explanation-finding, consolidation is the extension of the calculational precision of Newton's laws to explain the orbits of Jupiter's moons to unprecedented accuracy. In generalizing, it is calculating the electronic band structure of "whoofnium" after one has calculated that of whoofnium's cousins, "whifnium" and "whafnium" (Goudsmit, 1972).

- **Consolidation in the engineering method.** In function-finding, consolidation happens in the finding of new functions that extend previous functions, such as the need for a two-teraflop rather than just a one-teraflop computation rate. In form-finding, it is the steady Moore's law improvements in transistor density in integrated circuits—all these improvements required thousands of readjustments in the way the underlying materials, devices, and circuits were designed and fabricated. In exapting, consolidation can be seen in the use of existing forms for functions that are analogous to and extend previous functions.

Interestingly, perhaps because of their very different consequences, different nomenclatures have arisen for surprise and consolidation when they are downward-looking or upward-looking in the knowledge network. Looking downward (answer-finding), paradigm creation is sometimes described as "radical" change while paradigm extension is sometimes called "incremental" change. Looking upward (question-finding), paradigm creation is sometimes called "disruptive" change, while paradigm extension

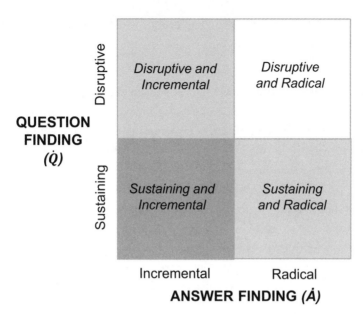

FIGURE 3-4. Nomenclature for four types of knowledge evolution, according to whether the evolution involves paradigm creation or extension, or involves looking upward (question-finding) or downward (answer-finding), in the seamless web of knowledge.

is sometimes described as change of a "sustaining" sort (Christensen & Raynor, 2013; Funk, 2013).

All combinations are possible. In the *lower left box,* sustaining change (looking upward with existing but improved questions) can be coupled with incremental change (looking downward with existing but improved answers), which represents paradigm extension looking upward *and* downward. In the *upper left box,* disruptive change (looking upward using brand-new questions) can be coupled with incremental change (looking downward with existing but improved answers), which represents paradigm creation looking upward but paradigm extension looking downward. In the *lower right box,* sustaining change (looking upward with existing but improved questions) can be coupled with radical change (looking downward for brand-new answers), which represents paradigm extension looking upward but paradigm creation looking downward. Finally, in the *upper right box,* disruptive change (looking upward to find brand-new questions) can be coupled with radical change (looking downward to find brand-new answers), which is paradigm creation looking upward *and* downward.

Surprise and Consolidation as Exploration and Exploitation

Why are surprise and consolidation both important to knowledge evolution? Because exploration and exploitation are both important as we create new, and make use of existing, knowledge to interact with the world around us. Surprise emphasizes exploration, the creation of new paradigms; consolidation emphasizes exploitation, the use and extension of existing paradigms. While both are important, the balance between exploration and exploitation, surprise versus consolidation, is not governed by a hard-and-fast rule. In an approximate way, the balance depends on the kind of world in which the evolving knowledge system is embedded and the speed with which evolution for survival must occur. The more complex and changing the world, the more reason to emphasize and incur the cost of exploration; the simpler and more static the world, the more reason to emphasize exploitation and avoid the cost of exploration.

In biological evolution, when organism variants are generated, those variants are tested in and by their world. Those that survive go on to reproduce, inheriting the original variation but also adding yet new variations. As formalized in Fisher's fundamental theorem of natural selection, the greater the variance in properties across organisms within each generation, the faster the rate at which the organismal population evolves, becoming fitter generation to generation. Variation, however, is costly, as most variants are less fit and die before reproducing, so the degree of variance is itself an optimizable and evolvable trait. The more complex and changing the world, the more reason to incur the cost of variance; the simpler and more static the world, the less reason to incur the cost of variance. The optimal rate of evolution or "evolvability" (Pigliucci, 2008) depends on the kind of world in which the organismal population is embedded.

In knowledge evolution, analogously, potential paradigms are generated, and those paradigm variants are tested by being played out in the real world. The measure of variance here is the degree to which the paradigm differs from or contradicts conventional wisdom, hence the degree to which one anticipates surprise. Thus, the paradigm generation process can be skewed either toward anticipated surprise or anticipated consolidation. Skewing toward surprise, however, is costly, as most potential paradigms that disagree with conventional wisdom are wrong and will have low utility. Thus, the optimal degree of variance depends on the kind of world in which the cognitive entity is embedded. At one extreme, if the world is complex or changing rapidly, then the optimal variance might weight anticipated surprise more heavily. Indeed, at this extreme, it might be optimal to explore new paradigms simply for their novelty and potential for surprise (Stanley & Lehman, 2015). At the other extreme, if the world is simple or changing slowly, then the optimal variance might weight anticipated consolidation more heavily. Why not make use of conventional wisdom rather than make a risky attempt to overturn it?

Thus, there is a balance between paradigm creation and extension, but the precise balance is situational. Human societies and organizations might adopt the balance appropriate for worlds to which they had to adapt during the long-term course of human evolution. An engineered or augmented human cognition might adopt a balance more appropriate to the current world, and a purely artificial cognition might adopt whatever balance is appropriate to the world into which humans have embedded it. Most importantly, a human society with sufficient self-understanding might adopt a balance that is optimal for its environment and then might implement that balance in public policy that determines relative investment in the two—as discussed in Chapter 4, relative investment in research versus development.

3.2 Surprise Spawns Consolidation Spawns Surprise

Surprise and consolidation are not just part of the natural rhythm of punctuated equilibria in knowledge evolution. They are part of a natural and holistic feedback cycle in which each spawns the other. The creation of a new paradigm naturally creates open space for the new paradigm to be extended and consolidated—to strengthen, and be strengthened by, the knowledge on which it draws. The extension of a paradigm, in turn, sows the seeds for the creation of yet new paradigms. As a paradigm strengthens knowledge modules, knowledge modules improve. These improvements may be individually small, but they accumulate, ultimately crossing performance thresholds that enable completely new questions to be asked and answered; in turn, enabling the creation of new paradigms elsewhere in the network. As a paradigm saturates in performance, on occasion even reaching a dead end, it can create pressure for new paradigms to take its place or for new subparadigms to take the place of subparadigms on which it relies.

Surprise Spawns Consolidation

Surprise and paradigm creation are singular events that create open space and possibilities for consolidation and further development in directions different from those occupied by previous paradigms. This open space is created because paradigm creation is accompanied by a dual, simultaneously occurring process of destruction that removes tethers and limitations imposed by previous ways of thinking. The two occurring together are akin to what Joseph Schumpeter called "creative destruction" (Schumpeter, 1942, p. 83):

> The opening up of new markets, foreign or domestic, and the organizational development from the craft shop and factory to such concerns as U.S. Steel illustrate the same process of industrial mutation—if I may use that biological term—that incessantly revolutionizes the economic structure from within, incessantly destroying the old one, incessantly creating a new one. This process of Creative Destruction is the essential fact about capitalism. It is what capitalism consists in and what every capitalist concern has got to live in.

Creative destruction sweeps out old ways of thinking and doing, replacing them with new ways, not only with greater power but also with different potential. The theory of special relativity, in which time dilates with velocity, supersedes the notion that time is immutable for all objects regardless of their velocity, with completely different ramifications for ultra-accurate measurement of time. The telephone superseded the telegraph while adding voice communication. The iPhone has the potential to supersede the landline phone completely, while also adding new computing and Internet functionality not possible with a landline phone.

In a sense, creative destruction applied to paradigms is a Gestalt-like event. Seeing the new paradigm renders old paradigms invisible. Much like the illustration of the old woman / young woman shown in Figure 3-5, in which one can see either the old or young woman, but not both simultaneously, one cannot interpret mass and motion in both Newton's and

FIGURE 3-5. Young woman / old woman Gestalt, seen either as the old woman or young woman, but not as both simultaneously. *Credit:* W. E. Hill, "My Wife and My Mother-in-Law" / Wikimedia Commons.

Ptolemy's paradigms simultaneously. Switching from one interpretation to the other takes time and effort. Once the automobile is viewed as the natural means of transportation, one no longer thinks of horses when one thinks to transport oneself—horses become invisible in the context of transportation. Indeed, the degree to which a new contribution to human knowledge is "paradigm changing" can be thought of as the degree to which it disrupts and makes previous paradigms invisible.

New paradigms are, in this sense, *incommensurate* with old paradigms, with paradigm creation opening completely new open space for paradigm extension. Speciation spawns adaptation in evolutionary biology, creating a new environmental niche that the new species occupies and to which it adapts. Paradigm creation, like speciation, is a discontinuous punctuation event in which a subpopulation (in this case, a knowledge module) becomes separated from its larger population and thrust into a new environment; paradigm extension is like the subsequent evolutionary path of the subpopulation as it gradually adapts to its new environment in the hierarchy. As articulated by Ron Adner and Daniel Levinthal (Adner & Levinthal, 2002 , pp. 50–66):

> Thus, a technology undergoes a process of evolutionary development within a given domain of application. At some juncture, that technology, or possibly set of technologies, may be applied to a new domain of application. The technological shift necessitated by this event is modest. Just as biological speciation is not a genetic revolution—the DNA of the organism doesn't suddenly mutate—technological speciation is not usually the result of a sudden technological revolution. The revolution is in the shift of application domain. The distinct selection criteria and new resources available in the new application domain can result in a technology quite distinct from its technological lineage. Framing technology evolution in terms of speciation leads us to differentiate between a technology's technical development and a technology's market application.

Ultimately, new paradigms create new knowledge modules that can also make irrelevant older knowledge modules created by older paradigms. Creative destruction applies not only to the metaknowledge of paradigms, but also to knowledge itself. New knowledge modules can supersede and make less relevant (and sometimes irrelevant) old knowledge modules. The new knowledge module could be scientific: a heliocentric rather than a geocentric world for explaining astrophysical observations. The knowledge module could also be technological: transistors that make irrelevant vacuum tubes for many electronic switching applications. New paradigms are the exemplar metaknowledge that created these new knowledge modules to begin with; thus new metaknowledge and new knowledge are created simultaneously, and old metaknowledge and old knowledge are destroyed simultaneously.

Consolidation Spawns Surprise

After the punctuation event, after the surprise and creation of a new paradigm, the paradigm's "equilibrium" period begins with consolidation and extension. The knowledge modules on which paradigms act are typically strengthened and improved, and, in some cases, ultimately reach the limits of their strengthening and improvement. Because knowledge modules are interconnected throughout the seamless web of knowledge discussed in Chapter 2, improvement or absence of improvement in one knowledge module has ripple effects throughout the web of knowledge. Sometimes those ripple effects are small and continuous, but on occasion they spawn surprise and the creation of new paradigms. This kind of surprise can emerge looking "upward," "downward," and "sideways" in the hierarchy of knowledge.

Looking Upward

Looking upward in the network of knowledge, surprise often emerges from simple accumulation of improvement. Such accumulation is often "geometric" and not just "arithmetic" (Farmer & Lafond, 2016). As articulated by Jeff Funk (Funk, 2013, p. 41):

> Geometrical scaling is one type of technological trajectory within a technology paradigm, and some technologies benefit more from it than do others. Those that benefit over a broad range of either larger or smaller scale typically have more potential for improvements in cost and performance than do technologies that do not benefit at all or that benefit only over a narrow range. Technologies that benefit from increases in scale do so because output is roughly proportional to one dimension (e.g., length cubed, or volume) more than are costs (e.g., length squared, or area), thus causing output to rise faster than costs as scale is increased. For technologies that benefit from reductions in scale, they do so because reductions in scale lead to both increases in performance and reductions in cost and the benefits can be particularly significant. For example, placing more transistors or magnetic or optical storage regions in a certain area increases speed and functionality and

reduces both power consumption and size of the final product. These are typically considered improvements in performance for most electronic products; they also lead to lower material, equipment, and transportation costs. The combination of increased performance and reduced costs as size is reduced has led to exponential changes in the performance-to-cost ratio of many electronic modules.

Even when the accumulation of improvement is slow, the ripple effect upward can be profound. After a long period of cumulative improvement, even if initially slow hence not readily apparent, the magic of "compound growth" can enable scientific explanation or technological form to cross a threshold of performance beyond which completely new questions can be asked or answered. Nature is full of threshold effects, where the crossing of some threshold of performance reveals completely new phenomena or enables completely new applications. In science, for example, the emergence and development of quantum mechanics enabled completely new ways of answering questions in chemistry, and the emergence and development of the theory of evolution enabled completely new approaches to answering questions in biology. In the use of technology to make observations and find stylized facts, it is common for technologies to cross thresholds where they begin to reveal new phenomena on new space and time scales. Existing science is then often found inadequate, motivating new science. In the use of technology to directly fulfill human function, it is common for technologies to cross thresholds where previously unfulfillable functions can suddenly become fulfilled. An iPhone with a just-low-enough power consumption to enable single-day use without requiring a battery recharge is transformative for people's lifestyles. A human who can just migrate across a savannah is transformative for the human race. Radial tires, initially developed for specialty markets after having sufficiently reduced cost and increased longevity, entered the bias-ply tire market, eventually displaced bias-ply tires, and became mainstream (Adner & Levinthal, 2002). The advances in the mid-2010s in deep learning were largely not algorithmic in nature; they were due to the cumulative advances in computation power and data availability associated with underlying technologies, advances that enabled deep learning to cross the threshold for human-like performance in image recognition.

All these examples can be thought of as generalization in science, and as exaptation in technology: the sudden onset of a new question at a higher level in the knowledge network that can be answered as a result of a gradual improvement in knowledge one level below in the knowledge network. Paradigm extension at a lower level of the network leads to paradigm creation at a higher level of the network: adaptation begets exaptation.

Interestingly, the accumulation of improvement, the ability to make desired improvements, is often taken for granted. While paradigms are being extended and becoming more powerful, they can give an impression of confidence, almost overconfidence, that all questions will someday be answered within the paradigm. In mechanics, Newton's laws certainly gave that sense; in electronics, the CMOS (complementary metal oxide semiconductor) integrated circuit certainly gave that sense. This can lead to the erroneous notion that the supply of answers is virtually infinite and, hence, what answers are found is mostly determined by the demand for which questions to answer—that the supply of inventions is unlimited and simply reflects the allocation of resources to respond to the demand (Schmookler, 1966). In fact, at any instant, existing paradigms form an interlocking frontier of possibility, and as these paradigms become exhausted, that frontier of possibility increasingly constrains the types of demand that are possible to supply (Rosenberg, 1974). In other words, paradigm exhaustion limits the ability of technology to meet perceived demand—the ability to provide answers to known questions. In such situations, neither demand subsidies for the cost of the product, nor supply subsidies for development efforts to improve the product, will lead to significant progress. Paradigm exhaustion, or the appearance of paradigm exhaustion, however, does contain the seeds and driving force for paradigm creation—both at the same level and at lower levels of the knowledge network.

Looking Sideways

Looking at the same level in the knowledge network, a new paradigm might simply replace an old paradigm—explaining similar scientific facts or fulfilling similar technological functions but with greater power.

In science, as scientific paradigms are extended and gain in predictive power, even small deviations from predictions become apparent.

These deviations represent anomalies that must be explained, sometimes by continued elaboration of the paradigm but sometimes (as when classical mechanics was superseded by quantum mechanics), by the creation of an entirely new paradigm.

In technology, as technological paradigms are extended and exhausted, they can likewise spawn new paradigms. Brian Arthur explains this natural cycle (Arthur, 2009, p. 141) as the

> origination of a new principle, structural deepening, lock-in, and adaptive stretch. . . . A new principle arrives, begins development, runs into limitations, and its structure elaborates. The surrounding structure and professional familiarity lock in the principle and its base technology. New purposes and changed circumstances arise and they are accommodated by stretching the locked-in technology. Further elaboration takes place. Eventually the old principle, now highly elaborated, is strained beyond its limits and gives way to a new one. The new base principle is simpler, but in due course it becomes elaborated itself. Thus, the cycle repeats, with bursts of simplicity at times cutting through growing elaboration. Elaboration and simplicity alternate in a slow back and forth dance, with elaboration usually gaining the edge over time.

Looking Downward

Looking downward in the knowledge network, surprise might emerge from submodules below: when the performance of knowledge modules at one level is limited by a particular submodule, enormous pressure can be placed on improving that submodule.

Whenever Moore's law improvements in integrated circuit chip scaling have appeared to be slowing, tremendous pressure was placed on improving the most critical submodules. Many of the new process steps associated with the continuation of Moore's law were new submodules that were surprising to conventional wisdom at the time. Copper, despite the known deleterious effects of copper on silicon semiconductor properties, took the place of aluminum for metallization. Chemomechanical polishing, despite its worrisome dirtiness, came to be used for creating ultrasmooth surfaces to facilitate subsequent processing. Likewise, high-dielectric-constant dielectrics, despite their process complexity, came to be used for reducing the thickness of gate insulators in metal-oxide-semiconductor field-effect transistors. These new submodules and surprises gave the appearance of steady extension of Moore's law and of "business as usual," but business one level below Moore's law was far from usual.

3.3 Cycles of Surprise and Consolidation: Artificial Lighting

Because surprise spawns consolidation and consolidation spawns surprise, surprise and consolidation occur in cycles. Here, we discuss examples of such surprise and consolidation in technologies for artificial lighting. We choose artificial lighting for three reasons.

First, light of any kind has been immensely important to humans and human society. Natural light from the sun and the visual information it carries after interaction with the environment, is central to human function, as evidenced by the highly evolved and exquisite human eye as an optical instrument, by the large fraction of the human brain devoted to visual signal processing, and by our extreme dependence on vision technologies, such as eyeglasses. Artificial light from lighting technologies expands human function so significantly—into the night and into enclosed spaces during the day—that the darkness we experience in its total absence often evokes complete disorientation, if not terror. Indeed, the sheer economic demand for artificial lighting has been a major driver for the large-scale adoption of various fuels over the course of human history (Weisbuch, 2018).

Second, the interface between (the protocol that connects) lighting technology and lighting use is relatively simple. Despite many complex characteristics (safety, reliability, cleanliness, convenience, color rendering quality, and directability) that might determine desirability, for many use cases one characteristic has been decisive: cost of light in dollars per megalumen hour ($ / Mlmh), where a lumen is a measure of the amount of visible light as perceived by the human eye. With this simplification, lighting approximates the hourglass of Chapter 2, in which a well-defined protocol screens the internal workings of the lighting technologies from the myriad uses for those same technologies. If a new technology produces light at a lower cost, it stands a good chance of supplanting the older technology regardless of how it will be used; and if a new use for light arises, it stands a good chance of being able to use whichever technology produces light at the lowest cost.

Third, lighting exemplifies surprise and consolidation, having gone through successive paradigms: candles and oil, gas, electric filament, and, most recently, electric solid-state lighting. As a result of the successive creation and extension of these paradigms, the cost of light decreased from ~15,000 \$ / Mlmh in 1800 to ~2 \$ / Mlmh in 2000, an astonishing factor of ~7,500 (Figure 3-6). At the same time, the consumption of light increased from ~0.0027 Mlmh / (person-year) in 1800 to ~140 Mlmh / (person-year) in 2000, an even more astonishing factor of ~50,000 (Tsao & Waide, 2010). These massive improvements in lighting technology and increases in lighting use were mediated by cycles of surprise and consolidation.

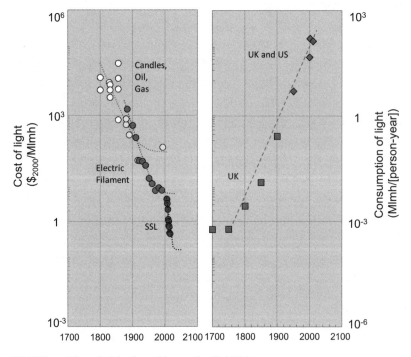

FIGURE 3-6. Three-hundred-year history of artificial light.

Oil Lighting

Artificial lighting did not begin with the modern human; it began with our protohuman ancestors who first domesticated fire beginning at least five hundred thousand years ago. But by at least fifteen thousand years ago, modern humans were burning fat in stone lamps with shallow depressions for the melted fuel (Nordhaus, 1996). For our purposes, we call this genre of lamp and its successors "oil lamps," as they are based on hydrocarbons that are liquid at their operating temperature. Perhaps the two most important surprises created in this genre of lighting had to do with the geometry of combustion: the bringing together of liquid fuel and oxygen from the air to create the fire that leads to light.

The first surprise was the wick. Invented in ~3000 BCE, the wick transported the liquid fuel upward by capillary action from the fuel source at the bottom of the wick to the site of fuel combustion at the top of the wick. The top of the wick being more fully surrounded by air, and, delivery of sufficient oxygen being the previous limiting factor in combustion, the fuel could be burned more efficiently. The simple wick was as revolutionary in the development of artificial lighting as was the wheel in the history of transport. Its invention occurred too early in pre-history for its story to be known, but one can easily imagine how surprising it would have been to find that the partial immersion of a fibrous string in oil would facilitate a vastly more efficient and controlled burning of that oil. This invention was form-finding of the highest order: the known question of providing illumination was answered in a completely new way with wicked oil lamps. Moreover, that new way was found almost certainly either serendipitously or through purposeful tinkering rather than through scientific insight—it was only thousands of years later that a scientific understanding of how oil lamps "work" was pieced together by Michael Faraday (Hammack & DeCoste, 2016).

The second surprise in the history of oil lighting was the Argand lamp. Invented in 1781 by the Swiss chemist Aime Argand, the Argand lamp had three striking innovations: a hollow tubular wick that enabled air supply to reach the wick from the inside *and* the outside, a glass cylindrical chimney to protect and facilitate convection of that air, and a mechanism for varying the length of the wick and thus regulating the ratio of oil to oxygen and tuning the intensity of the light. Like the wick, the Argand lamp was an unlikely invention, but also one complex enough that it could perhaps only have come about through the emerging science of the time. Argand had studied with Lavoisier, who in the 1770s had discovered oxygen as a component of air and as a necessary ingredient for combustion. That just-born scientific knowledge provided a new answer, "combustion requires oxygen," to the old question: "How do we explain combustion?" And that question-and-answer pair, once generalized to "the more oxygen, the better the combustion" led to an answer to a completely different question: "How can we provide brighter illumination?" This invention was deep into the implausible and next-adjacent possible and exemplified the use of science for engineering invention. Ultimately, the Argand burner became "to the nineteenth-century household what the electric light bulb is to the twentieth century" (Schivelbusch, 1995, p. 14).

These two surprises immediately and considerably improved oil lamp efficiency: as we now know, wicked lamps are better controlled and at least five times more efficient than non-wicked lamps, while Argand lamps are seven to ten times more efficient than simple candles, and they generate less smoke. In addition, the resulting new paradigms were compatible with significant subsequent consolidation and extension. Among those consolidations and extensions were myriad large and small improvements in wick materials, in the oils used to fuel the lamps, and in the overall lamp architectures that support the wick and the oil. By and large, these improvements made use of the wick and the fuel as the prime "submodules," whose interaction protocol was simply that the first must be able to wick upward the second. So long as that protocol was obeyed, each could improve separately.

Paradigm extension in oil lighting—the steady continued decrease in cost of light enabled by the steady evolution of wicks, oils, and lamp architectures—was thus just as important as the paradigm shifts themselves. Moreover, as the cost of light passed below various thresholds, new uses for lighting became economical, leading to exaptation of the improved technology to those uses. For example, for thousands of years, artificial lighting was mostly confined to residences for private indoor lighting. Though artificial lighting was also used by individuals to navigate the outdoors at night for private outdoor lighting, it was not used for street lighting or other types of public outdoor lighting. As oil lighting technology continued its steady improvement, the cost of light reached a threshold low enough to enable public outdoor lighting. Later, artificial lighting began to be used more widely in retail shops, creating new paradigms for retail: shops transformed from dingy warehouses into more colorful showrooms. Thus, the increasingly widespread use of lighting changed the very rhythm of daily life as activities, which could be shifted later and later in the day. In other words, paradigm shifts cascaded upward from lighting technology, to lighting use, to human lifestyle.

Gas Lighting

Even as oil lighting continued to evolve in the eighteenth and nineteenth centuries, most notably with the Argand lamp and with kerosene fuel from petroleum, it was far from perfect. It was not as bright as desired for many uses, it flickered and was highly nonuniform in time and space, and lamp fuel reservoirs had to be refilled often. Moreover, the standard oil lighting paradigm was, in fact, exhausting its technical performance. In this paradigm, the mechanism for the conversion of fuel energy to lighting relied on combustion being incomplete, with soot being formed and then heated to incandescence. Incomplete combustion meant an inherent inefficiency in the lamp, an inefficiency that could only be circumvented by a new principle of operation. Oil lighting was ripe for a shift to gas lighting, and this came in the form of two surprises.

The first surprise was the invention of central distribution of gas for lighting in 1792 by William Murdoch, the progenitor of gas lighting. Centrally distributed gas for lighting had many benefits: it was bright, it was uniform, and, most surprising, it could be economical.

The second surprise was the incandescent mantle, a device used to create bright white light when heated. Invented in 1885 by Carl Auer von Welsbach, a brilliant Austrian chemist, the mantle made practical an entirely new mechanism for luminescence (Weisbuch, 2018). In previous mechanisms, combustion, though incomplete, had played two roles: first, it created soot (carbon particles); second, it heated the soot to blackbody incandescence. In the new mechanism, combustion heated a prefabricated mesh of rare-earth-doped oxide to incandescence. Soot was no longer necessary, so combustion could be more complete. This new mechanism had been studied previously, even by Welsbach himself, but had been dismissed as impractical, due to the inability to mass-fabricate low-thermal-mass meshes that could be reliably heated to incandescence. The breakthrough came as a total surprise. While Welsbach was pursuing laboratory studies on rare earths, a solution of rare earth salts accidentally boiled over and evaporated on fibers at the ragged edge of an asbestos card used to support the beaker that contained the solution. Welsbach noticed that the flame playing around the card caused the salts to luminesce, and he immediately understood the significance—another case of chance favoring the well-prepared mind. Numerous experiments ensued that ultimately led to practical gas mantle lamps and revolutionized gas lighting.

As had happened with oil lighting, the surprise in gas lighting enabled immediate performance and efficiency improvements, which were compatible with subsequent performance and efficiency improvements as the new paradigm was consolidated and extended. Among those consolidations and extensions were myriad large and small improvements in the gas burners, the gases used to fuel the lamps, and the mantles. And, ultimately, because mantles were compatible with both oil and gas lighting, and because high-vapor-pressure oils could be readily turned into gases, a hybrid oil / gas lighting technology was developed. Pressurized kerosene lamps (such as Coleman lamps, still popular) rely on a hand pump to pressurize the liquid fuel, forcing it from a reservoir into a gas generator; vapor from the gas generator burns, heating a mantle to incandescence, and regenerating some of the heat to the gas generator.

Paradigm consolidation and extension associated with gas lighting (particularly the steady decrease in cost of light) were thus just as important as the paradigm shifts themselves. And, as cost of light decreased, new uses for lighting became economical, which led to exaptation of the technology to those uses. The first large-scale use of gas lighting came even well before the invention of the incandescent mantle—the factory. These were the early days of the industrial revolution, when factories demanding light were popping up *en masse*. Particularly in England, where coal and coal gas were plentiful, gas lighting became ubiquitous (Schivelbusch, 1995). The use of gas lighting for public streets was not far behind. By 1816, most streets in London were lit by gas; by the mid-1820s, most of the big cities were supplied with gas; and by the late 1840s, gas lighting was dominant in the cities and had begun to penetrate small towns and villages (Schivelbusch, 1995, p. 110).

Electric Lighting

Even as gas lighting continued to evolve in the nineteenth century, most notably with central gas distribution and the incandescent mantle, it had shortcomings. Gas lighting consumed oxygen, a major problem in closed indoor spaces; it generated much more heat than visible light; incomplete combustion and trace amounts of ammonia and sulfur blackened and caused damage to ceilings, surfaces, and paintings; and gas plumbing was expensive and leaks dangerous. These imperfections, combined with the sheer usefulness of light, motivated development of alternatives both to gas and oil lighting, still in use. Gas and oil lighting were ripe for replacement by electric lighting, and this came in the creation of two more surprises.

The first surprise was the durable carbon filament lamp. This new lamp involved many efforts and subefforts competing and cooperating with each other. Up to the instant of success there was considerable uncertainty as to whether it could be made practically durable. When the prototype lamp was finally demonstrated, it was recognized immediately as superior to any contemporary light source. Ten times brighter than gas mantles and one hundred times brighter than candles, its immense utility was immediately visible. The problem it intended to solve was clear: the demand for more durable, brighter light sources. But the solution to the problem up until the instant of success was much less clear.

The second surprise in electric lighting development was central generation, transmission, and distribution of electricity. Indeed, Edison succeeded in part because he realized that the goal was not to create the first reliable light bulb, but an entire practical commercial system of electric lighting—reliable electricity generation, transmission, and distribution *combined* with conversion into light. And, though lighting was the initial *raison d'être* of central generation, transmission, and distribution of electricity; because of the generality and ease of use of electricity for myriad other purposes, the exaptive spillover of such central generation, transmission, and distribution was enormous.

Consider the audacity of these two paradigm shifts happening almost simultaneously, with the success of each requiring the other's success. As with the Chapter 2 example of the iPhone, the two paradigms were jointly in the next-adjacent possible and not in the possible or even adjacent possible.

Just as with the surprises in oil and gas lighting, these surprises in electric lighting enabled immediate performance and efficiency improvement and were compatible with subsequent performance and efficiency improvement as the paradigms were consolidated and extended. Among those paradigm consolidations and extensions were myriad large and small improvements in both electric filament lamps and in electricity generation, transmission, and distribution systems. By and large, just as discussed above for oil lighting, these improvements made use of the lamps and the electricity generation, transmission, and distribution systems as separate "submodules," whose protocol is simply that the first must be able to be powered by the second. As long as that protocol is obeyed, each could improve separately. The protocol instantiation was the so-called E26 / 27 Edison socket, with its right-handed twenty-six- or twenty-seven-millimeter-diameter mechanical and 120VAC (in North America) or 230VAC (in Europe) voltage standards. That socket dominated electric lighting from its inception in 1909 until even now, in 2021 more than a hundred years later. The Edison socket played the role of the hourglass discussed in Chapter 2. It was a fixed protocol that enabled the electric filament lamp and the electricity generation, transmission, and distribution system to evolve as independent submodules, so long as they both "obeyed" the protocol.

Just as paradigm shifts enabled advancements in oil, gas, and electric lighting, subsequent paradigm consolidation and extension were equally important. A steady continued decrease in cost of light was enabled by the numerous ways in which electric lamps and the electricity generation, transmission, and distribution system evolved. In 1883, Edison marketed the first electric light at a price equivalent to gas lighting in town, close to the price of kerosene lighting. In the ensuing hundred years, as seen in Figure 3-6, prices decreased by nearly one hundred times. And, again as with oil and gas lighting, as the cost of light decreased below various thresholds new uses for lighting were enabled, which led to exaptation of the improved technology to those uses, with added benefits such as zero pollutants and no explosion hazard. Indeed, the penetration of electric lighting into virtually all uses seems now, in hindsight, inevitable: by 1925, half of all homes in the United States had electric power and lights (Bowers & Anastas, 1998).

Solid-State Lighting

Even as electric lighting continued to evolve in the first half of the twentieth century, the electric filament lamp was still relatively inefficient hence costly, and electric discharge lamps were complex, costly, and had poor color rendering quality. These and other imperfections, combined with the sheer usefulness of light, motivated development of alternatives to electric filament and its cousin, discharge lighting.

Coincidentally, the twentieth century also saw advances in the science and technology of semiconductors. The advances were highly interactive, transforming both our scientific understanding of the universe and the technologies by which we live our daily lives. The one semiconductor materials system that had remained elusive through the years of intense development of Si and the "conventional" III-V materials, however, were the "unconventional" III-N semiconductor materials—the materials that had wide enough electronic bandgaps to be able to emit visible light and, thus, serve as a foundation for semiconductor-based lighting. Between 1965 and 1970, RCA established a plausibility proof that some blue light could be "coerced" out of gallium nitride (GaN). But the obstacles were so great that, by the 1980s, most researchers had abandoned the field. As lamented by Hiroshi Amano and Shuji Nakamura in their Nobel lectures much later, single-digit attendance at the GaN sessions at technical conferences was routine. The pivotal surprise was the bright blue light-emitting diode (LED) in 1994. During that year, Shuji Nakamura began to publish on record-setting high-efficiency blue-light-emitting indium gallium nitride (InGaN) LEDs, and he began to give what were then described as "rock-star presentations," mesmerizing standing-room-only crowds of thousands of conference attendees with the blindingly bright LEDs he brought with him (Tsao et al., 2015).

These demonstrations shocked the science and technology community—going completely against then-prevailing scientific understanding and conventional wisdom, and an extreme example of disconfirming disbelief. Isamu Akasaki, Hiroshi Amano, and Shuji Nakamura, who shared the 2014 Nobel Prize in Physics for these breakthroughs, were guided more by their own intuition than by prevailing scientific understanding—and the scientific mechanisms underlying their breakthroughs ultimately took one to two more decades to painstakingly unravel.

Just as the new blue LED paradigm triggered efforts to explain it scientifically, it also triggered efforts to extend the technology. Though the prevailing science and technology community was, as mentioned, dismissive that the new paradigm was possible at all, once the new paradigm had been demonstrated, the community was primed to consolidate and extend it. High-quality large-diameter sapphire substrates could draw on three decades of progress in growth of large Si substrates for the Si-integrated circuit and other industries. Controlled InGaN LED epitaxial growth could draw on two decades of progress in precision epitaxy tools and processes. High-efficiency device and package designs could draw on three decades of progress in the "bandgap engineering" and sophisticated packaging used in virtually all modern III-V semiconductor devices. High-quantum-yield phosphors for converting blue into white light could draw on five decades of progress, beginning with phosphors for fluorescent lamps in the 1930s. The result was a rapid rate of reduction in cost of light, as can be seen in Figure 3-6, with a much steeper slope than any of the previous paradigms in lighting.

The 1994 blue LED, though striking, was born in a primitive state— very low efficiencies and very high cost. But as the cost of blue LED light gradually decreased, solid-state lighting was exapted to a stepping-stone series of uses through which the industry could grow and from whose revenue further research could be supported and further decreases in cost of light could be achieved. Among those uses were alphanumeric displays and indicator lamps, traffic signal lights, outdoor signage and displays, flashlights and camera flashes, backlights for liquid-crystal displays (LCDs), automotive lights, street lighting, and architectural lighting (Haitz & Tsao, 2011). All these "specialty" uses for what has now come to be called solid-state lighting (SSL) became economical at higher cost-of-light price points than general illumination, could be used to generate revenue and "protect" the emerging technology, and were essential to ultimately achieving the price points necessary for general illumination.

Within a decade and a half, by 2011, the life-ownership cost of lighting of SSL had decreased below that of electric filament lighting. In addition, SSL had other characteristics important to basic white-light illumination for humans: high color rendering quality and tailorable color temperatures, minimal or no flicker, long life, and negligible environmental and human toxicity. A massive transition to SSL in all general illumination uses began and is still in progress—in residences, factories, offices, and commercial spaces.

We can anticipate, interestingly, that these new uses will not proceed as quickly as the technology itself allows. Why? Because of lock-in. The infrastructure of lighting use developed for previous lighting technologies is locked in and prevents the free evolution of SSL technologies. As one simple example, the so-called E26 / 27 "Edison socket" for the standard light bulb is ubiquitous. By playing an hourglass role, it enabled independent innovation in the light bulb itself and in the electricity generation, transmission, and distribution system. But in the case of SSL, the socket also plays a deleterious lock-in role. SSL is a compact technology compatible with much smaller sockets, but the first wave of SSL are virtually all backward-compatible "retrofits" with the larger Edison socket. Only when a new socket standard evolves will SSL be free to unleash its full capabilities.

3.4 Recapitulation

To recapitulate, in this chapter we have described how the evolution of knowledge occurs via punctuated equilibria, in which periods of steady and relatively more gradual advance, that is, *consolidation*, are punctuated by surprising and relatively more sudden advances that we call *surprise*.

The mediators of that punctuated equilibria are paradigms: holistic combinations of knowledge "put to work" to advance technoscientific knowledge—a kind of metaknowledge of how to *use* knowledge modules to accomplish dynamic *changes* to knowledge modules. Knowledge modules are like nouns, both concrete and abstract. Paradigms are like sentences—they use knowledge modules to create or extend other knowledge modules. Unlike the knowledge modules they draw on, which are either scientific or technological but not both, paradigms are opportunistic, drawing on knowledge modules of either or both type(s) as appropriate for the task at hand.

Paradigms are created by the concomitant and discontinuous creation of new, unexpected, and *surprising* question-and-answer pairs. As outlined in Chapter 1, in 1944–1949, the new scientific understanding of the transistor effect and the new technology of the transistor device itself represented a new paradigm in semiconductor transistor engineering. Paradigm creation is accompanied by the destruction of previous paradigms, analogous to the process of "creative destruction," and involves both utility and surprise. For a new paradigm to be created and have lasting value, it must have utility. For a new paradigm to destroy an existing paradigm, it must be incompatible in some way with conventional wisdom, so there must be surprise. Thus, we can think of paradigm creation as "implausible utility." Along with this, as we will discuss in Chapter 4, comes another important facet of paradigm creation that we call "informed contrariness," when deeply informed individuals pursue directions contrary to conventional wisdom with a reasonable chance of overturning it.

Paradigms are also extended, and this we associate with consolidation: refinements of existing question-and-answer pairs that are more continuous and more predictable. Such paradigm extension is vital: the extension of the semiconductor transistor engineering paradigm over the past half century, along with the creation of many new subservient "subparadigms," has been essential to fulfill the potential of that paradigm. Paradigm extension is accompanied by a consolidation and strengthening of conventional wisdom. For paradigm extension to have lasting value, it must have utility. For paradigm extension to strengthen the current paradigm, it must be plausible to conventional wisdom, thus involving less surprise. Therefore, we can think of paradigm extension as "plausible utility."

Surprise and consolidation occur in an overarching feedback loop. On the one hand, surprise spawns consolidation: paradigms must first be created before they can be extended. On the other hand, consolidation spawns surprise: paradigms must be extended to achieve their fuller potential, and, as they are extended, they sow the seeds for the creation of new paradigms. They do this in two very different ways. In the first way, in response to existing questions they are trying to answer, improvements accumulate to the point of being able to be generalized or exapted to completely new questions. In the second way, also in response to existing questions, improvements become exhausted or threaten to become exhausted, leading to opportunities for them to be superseded by new and more powerful paradigms or for performance-limiting sub-paradigms below them to be superseded by new and more powerful subparadigms.

These cycles of surprise and consolidation can be seen from the history of artificial lighting. Human civilization has seen the successive creation of new lighting paradigms that surprised and upended conventional wisdom: from oil lighting, to gas lighting, to electric lighting, to, most recently, solid-state lighting. Each of these new paradigms was extended and made far more powerful, consolidating the new conventional wisdom before ultimately being replaced by the next paradigm.

4

Guiding Principles for Nurturing Research

In Chapters 1 to 3, we articulated our rethinking of the *nature* of research in which science and technology symbiotically coevolve, the finding of questions and answers occur in an intricate dance, and knowledge evolves by the punctuated equilibria of surprise and consolidation. In Chapter 4, we articulate our rethinking of the *nurturing* of research.

By focusing on research, the seeking of surprise and the overturning of conventional wisdom, we by no means intend to trivialize development, the consolidation and extension of conventional wisdom. Development is vital and necessary for the full power of existing paradigms and conventional wisdom to be realized. But research is also vital, and more fragile. In most R&D organizations, research is a "risk capital" effort with a budget much smaller than that of its development counterpart, thus easy to overlook. Because the outcomes of research cannot be scheduled or predicted in advance, they can neither be missed nor championed when they do not materialize. And, even when championed, research is more difficult to nurture than development. The tendency to reduce the uncertainty inherent in research is strong, but doing so essentially turns research into development (Narayanamurti & Tsao, 2018).

Even as we focus on research, we recognize that research and development share an intellectual synergy, and that development itself is vital to the success of research. By focusing on the overturning of conventional wisdom, we do not mean the indiscriminate overturning of *all* conventional wisdom. As discussed in Chapter 3, knowledge is woven together into a seamless web on whose shoulders it is usually advantageous to stand. Even as research surprises one piece of the seamless web, it makes use of the rest of the seamless web. Michelson and Morley's determination, as discussed in Chapter 2, of the constancy of the speed of light regardless of reference frame upended beliefs about the aether that supposedly filled space, but did not upend optical interferometry technology. Research thus naturally contains within it significant development. Even the smallest middle-of-the-night idea can, the next morning, launch a development-like miniproject requiring modification or extension of some physical instrument or calculational algorithm so as to test the idea. The miniproject might even turn into a major project requiring significant effort. The effort might even be so large (for example, a high-energy particle accelerator), as to require an entire organization such as CERN (the Conseil Européen pour la Recherche Nucléaire), the world's largest particle physics laboratory, to execute the project. Research cannot succeed without the support of development activities internal to it. But key to successful research is that its internal development activity take direction from research and be responsive to its opportunistic twists and turns. If the internal development activity, with its typically more conservative culture, becomes too powerful, research can become secondary to what the development organization does and wishes to do—the tail wagging the dog. Instead, the tail must be wagged by the dog.

In the remainder of this chapter, we outline a small number of guiding principles, illustrated in Figure 4-1, for organizations wishing to nurture research Although the principles were, in part, inspired by

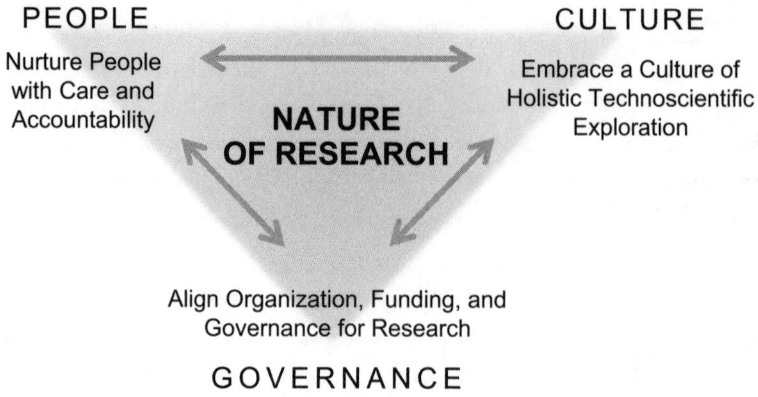

PEOPLE   CULTURE

Nurture People   Embrace a Culture of
with Care and   Holistic Technoscientific
Accountability   **NATURE**   Exploration
                          OF RESEARCH

Align Organization, Funding, and
Governance for Research

GOVERNANCE

FIGURE 4-1. Three guiding principles for nurturing research, related to governance, culture, and people.

corporate research laboratories such as Bell Labs, the principles are intended to be general enough to apply to a wide range of research organizations—from large industrial and government research laboratories to universities. Different parent organizations may have different overarching missions (universities to educate, industrial corporations to produce and sell goods and services, government institutions to provide particular public services), and these missions may be constraining in various ways. Different parent organizations may also have evolved particular organizational forms (university researchers with tenure, industrial corporations with Wall Street oversight, and government institutions with Office of Management and Budget oversight), and these may also be constraining in various ways. No matter what type of organization wishes to do and nurture research, however, it must follow similar guiding principles if it is to be successful.

Our first principle, *align organization, funding, and governance for research*, acknowledges that research is a highly specialized activity whose success requires organization, funding, and governance to be aligned with the peculiarities of that specialized activity. Our second principle, *embrace a culture of holistic technoscientific exploration*, acknowledges that technoscientific exploration is fragile and easily stunted; hence, there must be active nurturing of a culture that supports such exploration. Our third principle, *nurture people with care and accountability*, acknowledges that people are the beating heart of research—they, with all their diverse perspectives and idiosyncrasies, must be cared for, even as they are held accountable to high standards.

Throughout, we borrow liberally from a discussion of seven critical elements of research culture observed at the iconic Bell Labs (Narayanamurti & Odumosu, 2016, pp. 80–91): freedom to fail and the patience to succeed; collaboration as the primary mode of interaction; competition as the primary mode of individual aspiration; intense interactivity with peers; administrative leadership from within; egalitarian meritocracy; and excellence as a virtue in hiring, promotion, and review. However, instead of viewing these elements as observed facts about one unique research organization, here we view them as necessary characteristics that follow naturally from a set of guiding principles for the nurturing of effective research. Indeed, our emphasis on Bell Labs is not because these characteristics do not hold for other great industrial research laboratories (IBM, Xerox PARC, Dupont, GE) and research institutions (the Cavendish Laboratory, the Laboratory for Molecular Biology, Rockefeller University, Lawrence Berkeley National Laboratory, Janelia Research Campus)—it is because Bell Labs is particularly exemplary of these characteristics as well as deeply familiar to one of us (VN).

Note that these guiding principles are focused on how organizations can best nurture *research*. No organization, even one dedicated to research, nurtures *only* research. It must also nurture a wide range of other human physical and social needs, especially a sense of community and shared values. We assume that our guiding principles specific to nurturing research play out against such a larger backdrop human nurturing.

4.1 Align Organization, Funding, and Governance for Research

As a highly specialized activity, research must be treated differently from its equally important partner in knowledge production, development. Organization, funding, and governance must *align* with research, and here we discuss four important ways of bringing about such alignment. First, research funding should not be a casual investment. Research outcome cannot be predicted or guaranteed in advance, so research organizations should invest in research only if their purpose can accommodate the unexpected. Second, research organizations need to recognize that there is a deep difference in mindset between research, which seeks to surprise and overturn conventional wisdom, and development, which seeks to consolidate and strengthen conventional wisdom, necessitates that research be culturally insulated, though not intellectually isolated, from development. Third, to deal with the unanticipated, researchers and research leadership must be able to respond flexibly and opportunistically, and this requires block allocation of resources, both to research leadership at the organizational level and to researchers as people, not projects. Fourth, research leadership is critical. Research is not simply a matter of assembling researchers and giving them free reign—research must be orchestrated to maintain a delicate balance between organizational focus and individual freedom.

Fund Research to Achieve Metagoals, Not Just Goals

We start with a question: Why is research outcome difficult to anticipate? The answer: because research is defined by an "ultimate" metagoal to which lower-level "proximate" goals are subservient.

Just as for paradigms where metaknowledge makes use of but overarches knowledge, so metagoals make use of but overarch individual research goals. For example, in competitive sports an athlete's ultimate metagoal is to perform at the pinnacle of their sport. The athlete might set up all sorts of proximate goals, like controlling diet, weight training in particular ways, or responding in real time to competitors' behaviors in certain ways. But the athlete continually reevaluates these proximate goals according to whether they increase or decrease ultimate performance. Or, for example, in biological evolution an organism's ultimate metagoal is to survive and reproduce. The organism might have all sorts of proximate goals, like escaping a predator, capturing prey, or attracting a mate. But these proximate goals continually adapt to the selection pressure imposed by the overarching metagoal.

For research, the ultimate metagoal is the surprise discussed in Chapter 3: surprise that changes the way we think and do and that leads to paradigm creation and destruction. *En route* to this ultimate metagoal, research can and must have proximate goals having to do with the intellectual work at hand. We go to our instruments with the intent of making particular observations, we puzzle over our observations with the intent to explain, we redesign our device with the intent of changing its performance, and we puzzle over the change we actually observe. But when we uncover something unexpected, when our actions and thoughts are led in a different direction (perhaps because we have found a solution to an unexpected and more important problem or found an unexpected solution to an existing problem), we must not block off the new direction simply because it differs from our initial course. If research finds, *en route* to its proximate goals, new proximate goals with potential for *more* surprise and *greater* creative destruction, then research must redirect its attention to those new proximate goals.

It is in this lack of adherence to proximate goals that research is less directed than development. Research flexibly accommodates opportunistic changes in proximate goals because, as we discussed in Chapter 2, research draws on the inherently unpredictable adjacent and especially next-adjacent possibles. What is possible to do or think depends on where one sits in the knowledge landscape. As one gains more knowledge and shifts to another position on the landscape, new actions and thoughts are revealed. Engaging in less directed research means being open to those new possibilities.

In the language of questions and answers (Chapter 2), we suggested that it is often in the finding of new questions that new possibilities open up; and in the language of the technoscientific method (Chapter 1), we noted that it is often in generalizing and exapting that new possibilities open up. As illustrated in Figure 4-2, in science, the constancy of c was an outstanding question; its answer was Einstein's theory of special relativity, and this answer then generalized in an unexpected way to the question of energy release on nuclear fission and fusion events. Also as illustrated in Figure 4-2, in technology, improved tools for spectroscopy was an outstanding question; one answer was the laser, whose narrow line widths enabled unprecedented accuracy in the determination of energy levels in atoms and molecules, and this answer then was exapted in a totally unexpected way to the question of how to achieve higher bandwidth and longer-distance communication beyond what was possible with electric signals and copper transmission lines.

These are both "big" examples in terms of breadth of knowledge domains affected, but similar examples of the opportunistic finding of new questions that lie outside the knowledge domain of initial interest abound at all scales. At any scale, it takes only a few iterations of

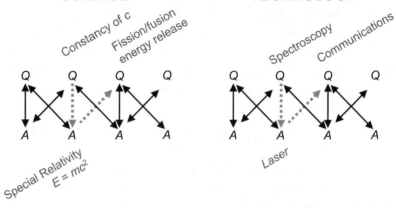

FIGURE 4-2. Examples of unanticipated research outcomes in science and technology.

question-finding for research impact to shift outside the knowledge domain of initial interest. Indeed, the intellectual impact of research is overwhelmingly likely *not* to be in the precise knowledge domain of initial interest. As articulated by Herb Kroemer, corecipient of the 2000 Nobel Prize in Physics for developing semiconductor heterostructures used in high-speed and opto-electronics (Kroemer, 2001, p. 787):

> When I proposed to develop the technology for the DH [double heterostructure] laser, I was refused the resources to do so, on the grounds that "this device could not possibly have any practical applications," or words to that effect. By hindsight, it is of course obvious just how wrong this assessment was. It was really a classic case of judging a fundamentally new technology not by what new applications it might create, but merely by what it might do for already existing applications. This is extraordinarily shortsighted, but the problem is pervasive, as old as technology itself.... Any detailed look at history provides staggering evidence for what I have called ... the Lemma of New Technology: The principal applications of any sufficiently new and innovative technology always have been—and will continue to be— applications created by that technology.

Not only is the unanticipated the likely, indeed sought after, result of research, but the unanticipated is difficult for many organizations to derive business advantage from.

The difficulty derives in part because *all* new knowledge, whether inside or outside the knowledge domain of initial interest, is difficult to derive business advantage from. In the language of economics, knowledge is a nonrivalrous good—a good undiminished by consumption and whose use by one organization does not prevent its use by another organization. This public nature of knowledge is one of its most powerful strengths, of course, enabling it to grow cumulatively, standing on the shoulders of shared previous knowledge, as exemplified by today's open-source software community. But it is also one of its weaknesses. As with the tragedy of the commons, if everyone owns knowledge, then no one owns it. Economists have thus long argued that the resulting difficulty of deriving business advantage leads firms to underinvest in research (Arrow, 1962; Nelson, 1959).

Unanticipated new knowledge is even more difficult to derive business advantage from. The impact of unanticipated new knowledge spills over into unanticipated other knowledge domains (Stephan, 1996). Other firms who happen to be working in those other knowledge domains could easily derive greater benefit than the originating firm from the new knowledge. Moreover, the new knowledge even has the potential to "put out of business" parts of the larger institution in which the research organization is embedded, as new knowledge and technologies render the old obsolete. This is a negative externality of research that can cause other parts of the larger institution to be not only disinterested in but actively opposed to the research activity. As articulated by Otto Lin, president of the Industrial Technology Research Institute in Taiwan from 1988 to 1994 (O. C. C. Lin, 2018, p. 62):

> The management of disruptive innovation is a tricky undertaking and very dangerous if not handled properly. . . . As many existing units see it, supporting disruptive innovations that the R&D guys are working on is nurturing a force that, if successful, will become a terminator of its own business. What a prospect this will be! How enthusiastic will these business units be in supporting the new R&D program?

Thus, to benefit from new knowledge that lies outside the precise domain of initial interest—to count such knowledge spillover as success— the mission of the organization must transcend immediate or local business or economic profit; it must have a public goods component "bigger than oneself" (Mazzucato, 2011).

For a public or private research university whose mission includes the "advance of the frontiers of knowledge" (Summers, 2002), with the precise domain of knowledge left open-ended, unanticipated new knowledge of benefit to broader human society *is* fulfillment of the university's mission.

For a for-profit corporation, such a mission might derive from simple founder's idealism. This was the case with David Packard, the co-founder of Hewlett-Packard Labs (HP). He had an altruistic vision that HP should "return knowledge to the well of fundamental science from which HP had been withdrawing for so long" (Williams, 2008, p. 31). This was also the case with Alexander Graham Bell, the founder of AT&T, who would often invoke advancing knowledge as fulfilling a larger national calling to improve all of human society. It helped, of course, that, for more than half a century (1912–1984), AT&T was a regulated monopoly, protected from the business desire to benefit exclusively from the knowledge it created. AT&T's transistor was licensed for only a very small amount and for only a few years before it was released for free to the public domain (Braun et al., 1982).

For a not-for-profit government or philanthropic organization, such a mission might derive from a public service or philanthropic mandate, as with Sandia National Laboratories' metagoal "to render exceptional service in the national interest" or with the Janelia Research Campus' metagoal "to develop new technologies and apply them to challenging biomedical problems that may have a low probability of short-term success but a high potential impact" (Cech & Rubin, 2004, p. 1166).

Whatever its derivation, the organization's mission must provide overarching motivation for the creation of new knowledge that spills over outside the domain of initial interest. Otherwise, as discussed in the Introduction, the organization will face a situation in which it is unable to meet the metric of short-term private return-on-invested-capital imposed by Wall Street. Indeed, as also discussed in the Introduction, by the 1980s, large corporations (including AT&T, DuPont, General Electric, and Xerox PARC), and some publicly funded mission organizations had fundamentally reconsidered the role of research in their operations (Kingsley, 2004). Research became more tightly linked to the short-term business objectives of the corporation, essentially turning research into development. That said, we do not mean to say that research outcomes and impacts are antithetical to short-term business objectives of the corporation. Such impacts are more than welcome and should play, as discussed later in this chapter, a significant role in the organization's choice of knowledge domain within which to build research critical mass. But such impacts cannot be forced and must be secondary to the metagoal of surprise if research is to be true to its nature.

Inseparably linked with organizational mission is the source of funding for that mission. Corporate sources of funding increasingly cannot accommodate advance of knowledge for the long term and for the public good, due to the return-on-invested-capital pressures discussed above. Philanthropic sources of funding, however, *can* accommodate such advance in knowledge and are an important reason why research universities and institutes increasingly dominate the research landscape. Government sources can in principle accommodate such advance in knowledge, but in practice they increasingly do not. Government sources increasingly segregate their support of research into bins (scientific versus engineering research, so-called "basic" versus "applied" research), and increasingly seek the goal of utility in particular knowledge domains rather than the metagoal of surprise and surprise's inevitable spillover into broader knowledge domains.

Finally, we note that unanticipated new knowledge is not the only product of research organizations that requires a bigger-than-oneself mission. Researchers themselves are a product of the organization's research investments. Nurturing researchers from the earliest to the most productive stages of their careers is expensive. Anywhere along the way, even perhaps at the peak of their careers, their life situations may change, and they may decide to move on—sometimes to a different institution entirely. Such a move cannot be viewed as a failure of the organization. Researchers who have opportunities to move on and contribute to society in other ways are an important contribution of a research organization; indeed, they are a kind of yardstick by which to measure the organization's success at nurturing researchers. Thus, for this reason as well, the mission of the organization must transcend immediate or local business or economic profit.

Insulate, but Do Not Isolate, Research from Development

The unanticipated is not only the most likely but also the *desired* outcome of research. But the unanticipated can be uncomfortable, so there is a strong human tendency to avoid and prevent unanticipated outcomes (Narayanamurti & Tsao, 2018). The most common way is to treat research like development and to borrow development's most potent tool: the project and its machinery of scheduled milestones and deliverables. Moreover, it is easy to mistakenly believe that research is compatible with such scheduled milestones and deliverables. Even researchers themselves, when they formally communicate their results, as a matter of traditional practice typically conceal the twists and turns taken during the course of their research. The results of research are typically described after the fact in the most logically coherent way rather than as a story narrative. As articulated by Ernst Mach (Merton & Barber, 2004, p. 274):

> Euclid's system fascinated thinkers by its logical excellence, and its drawbacks were overlooked amid this admiration. Great inquirers, even in recent times, have been misled into following Euclid's example in the presentation of the results of their inquiries, and so into actually concealing their methods of investigation, to the great detriment of science.

In fact, research process *cannot* be scheduled or planned. Research begins with ideas and plans, but once contact is made with nature these initial ideas and plans will inevitably need to be changed. As the well-worn dictum states: "No plan survives contact with the enemy." Researchers must have the freedom to respond opportunistically as they explore, as they learn, and as they confront serendipitous events. In other words, research must not be "projectized."

How can the projectization of research be avoided? In organizations without significant development activities, such as universities and research institutes, projectization is less of an issue. But in organizations that devote significant time and energy to development activities, such as industrial corporations and mission-oriented government organizations, projectization can be a significant issue. Thus, care must be taken to insulate research organizationally from development so that each can create and maintain its own organizational culture—research a culture of research and development a culture of development. Only in this way can each specialize in what they do best—much like scout ants specialize in exploring independently for new food sources, while worker ants specialize in exploiting sources the scout ants have found. That said, the intellectual content of research and development organizations might overlap quite closely. Researching exotic new phenomena in semiconductor devices may build on similar fabrication, characterization, and modeling knowledge as developing the final process for the manufacture of a to-be-shipped-next-month semiconductor chip. Thus, though research and development are culturally dissimilar, and so must be culturally insulated from each other, they are often intellectually synergistic, and therefore should not be intellectually isolated from each other. Indeed, when the intellectual synergy between research and development is strong, it is tempting for leaders and managers to organize for intellectual synergy first, culture second. To illustrate this tendency, Figure 4-3 shows two possible structures for an organization that wishes to pursue both research and development in two knowledge domains: the synthesis and processing associated with a particular semiconductor material, GaN, and the physics and devices also associated with that material.

The structure on the left organizes groups by intellectual content, with researchers and developers residing in the same groups. Communication between researchers and developers is close, and intellectual synergy is high. Researchers can see developers' problems firsthand and can make use of these problems in their research. Developers can see new discoveries and inventions unfold firsthand and can develop them quickly. However, this structure leads to an inherent culture clash between

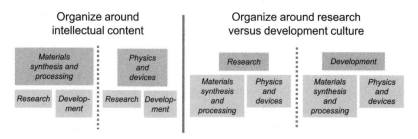

FIGURE 4-3. Two hypothetical organizational structures for research and development in a particular knowledge domain.

researchers and developers. Researchers seek to create conditions that maximize surprise; developers seek to create conditions that minimize surprise. If an odd observation is made under an unusual and accidental set of conditions, researchers might like to reproduce those conditions to understand what happened, while developers might want to ensure that those conditions never again occur.

The structure on the right instead uses culture as an organizing principle via separate research and development groups. Because the groups are organizationally separated, communication between researchers and developers may be less frequent. However, with this structure, distinct and internally consistent organizational cultures can be created and sustained to align with each culture's metagoals: researchers to explore the unknown and to maximize surprise, developers to exploit the known and maximize utility.

The second structure surpasses the first, by far. The reason: *we are humans first, intellects second.* Because we are humans first, we respond first and foremost to social cues, to local culture, to peer pressure, and to our organization's values and reward system. Only after we have adapted socially and culturally to our organization do we engage our intellects. For this reason, the "one company, one culture" mantra often heard today is deadly. When the research-development cultural boundary is too porous, development inevitably diffuses into research. Typically five to ten times larger than research, development outweighs research in sheer size. Moreover, development is more consistent than research with our human tendency not only to minimize uncertainty and surprise, but also to monetize our activity. Thus, when research and development cultures clash, development culture wins. And, once a research organization has drifted toward development, it is exceedingly difficult to shift it back toward research. Instead, it is best to nucleate a new organization— one whose leadership exerts strong effort to shape a culture aligned with the nature of research rather than one that inherits the culture of the parent organization.

All this said, we do not mean to suggest that research and development are in any way "at war" with each other. Both research and development are necessary to the advance of knowledge. We reject the primacy that research sometimes displays toward development, but we also reject the primacy that development sometimes displays over research.

Block-Fund People, Not Projects

Research success depends on more than its cultural insulation from development. It is also necessary to fund research differently. Because the fundamental units of development are projects, it is natural to fund development by project—and remunerate people for the projects in which they participate. The fundamental units of research, however, are not projects but people—research leadership and researchers themselves. It is people who are the stable entities through the complex and opportunistic coevolution of questions and answers as research progresses (Narayanamurti & Tsao, 2018). Thus, in research, it is people, not projects, that must be funded. Research organizations must be "block funded"—allocated resources at the organizational level that are managed with discretion and accountability by research leadership, with researchers being funded by research leadership as people, not simply as project participants. When research is thus organized, four important freedoms for research leadership and researchers result.

First, organizations experience the freedom to nurture and build capability, not just to harvest and make use of existing capability. Research leadership must at the organizational level have the freedom to recruit and hire researchers, thus building critical mass in new areas; to give researchers the opportunity to learn those new areas; and to develop new capabilities that may not pay off immediately but could be extremely valuable to the researchers over time. Just like an athlete must occasionally take time off from match play to hone new skills, researchers must occasionally step away from the laboratory to develop new capabilities. But when researchers are funded by project, a transactional "contractor" mentality prevails, in which projects wish to pay only for direct project work rather than for the building of capabilities—as captured by Larry Summers, former president of Harvard University, who said, "In the history of the world, no one has ever washed a rented car."

Second, research organizations are free to adapt to the unexpected. Projects have defined, specific goals with milestones, timelines, and budgets for getting to those goals. Research outcomes cannot, by definition, be anticipated, as research is guided by metagoals rather than fixed goals. Researchers must have the freedom to respond opportunistically to the unexpected; to pivot away from one investigatory course to another, if warranted; and to terminate one project plan in favor of a completely different project plan (Azoulay et al., 2011). Plans and projects are important, but in research, they must be flexible and cede primacy to people so they may change directions flexibly in service of research's overarching metagoal.

Third, researchers can be freed from writing detailed proposals, the precursors to projects in many of today's research environments. Project proposal processes are antithetical to research for several reasons. In today's research climate, proposals have a significant marketing component, hence funding often goes to the best-marketed proposal rather than to the best research. Because proposals are typically awarded based on peer review, where even one negative review can mean no award, proposals are written conservatively—to confirm rather than disconfirm conventional wisdom. Further, because proposals are usually written to do research now, rather than to build capability for doing research in the future, they disadvantage researchers who desire to move into new knowledge domains—researchers who have not already spent significant time in a new area will have very little chance of writing a credible proposal. Finally, because detailed proposals are time-consuming to write, they take enormous amounts of time away from research itself.

Fourth and finally, researchers become more free to collaborate or not collaborate as the situation warrants. Collaboration, of course, is essential to research, enabling researchers to capitalize on their strengths and compensate for their weaknesses. But collaborations also represent significant investments in time and energy, and because they are complicated by intellectual, cultural, and relational human elements, they can sometimes fail. Because researchers themselves have the most "skin in the game" and local knowledge, they must have the freedom to choose with whom to collaborate or not collaborate. Imagine Researchers A and B, newly aware of a potential and extremely productive collaboration between them. If both are already funded for projects, switching to the new idea requires them to break free from those projects and thus trade away their safe current funding. Or imagine Researcher A, asked by Researcher B to participate in an emerging project that is not, in her estimation, the best research to pursue, but one that she knows is more marketable. If Researcher A is not yet "covered," that is, if her own funding is uncertain, she is overwhelmingly likely to accept the project and the collaboration with Researcher B. When salary and time allotments are tied to projects, researchers do not have the psychological safety and incentives to make the best choices for allocating their collaborative time and effort.

Note that funding people, not projects, is not necessarily an either / or choice—hybrid choices exist in the gray in-between area. An increasingly popular hybrid funding model is the virtual organization that matrixes projects across people that reside in many different "line" organizations. These might be local "grand challenge" projects within an institution that draw on researchers from many internal suborganizations. These might even be national in scale, drawing on institutional partners all across the nation, as do the National Science Foundation's Engineering Research Centers and the Department of Energy's Energy Frontier Research Centers, Energy Innovation Hubs, and Quantum Information Science Centers.

Such virtual organizations do have some advantages. They can be established relatively easily and can quickly achieve critical mass and scale on big topics of known importance. They can draw strategically on interdisciplinary expertise spread over a wide range of institutions. They have finite lives, typically three to twelve years, so topic areas can evolve as learning takes place. But virtual organizations cannot substitute for line organizations, including those from which they draw their researchers and infrastructure. Virtual organizations harvest rather than nurture. When they are large enough, in principle they have the resources to nurture, but in practice they don't have the permanence or motivation for the deep engagement and mentoring needed to nurture researchers' careers.

Thus, research *line* organizations are critical. Only line organizations and their leadership are charged with nurturing and developing full human beings. Only line organizations know who just experienced a major change in life circumstance, who has been harboring a hobby interest in a side topic and would jump at an opportunity to engage in that topic more deeply, and who has been harboring a desire to shift away from research toward equally exciting and important development work. Only line organizations can provide deeper, nontransactional, and non-contractor-like nurturing of their researchers—doing for their people what recapitalization does for physical infrastructure. And line organizations at scale, such as industrial corporations and national labs, can be especially important, particularly for knowledge domains believed to have long-term opportunity for discovery and invention. Such organizations can combine critical mass and scale *with* the long-term nurturing that is difficult for matrixed virtual organizations tied to shorter time horizons.

Leadership

Organizational purpose, structure, and resources—all these, as we have just discussed, must be aligned with research. Someone, though, must construct and oversee that alignment, and that is the essential role of leadership. Only leadership can set the purpose of the organization, one that must include a "bigger than oneself" mission. Only leadership can align organizational structure so that research is culturally insulated, but not intellectually isolated, from development—so that a research culture can thrive and not become secondary to development. Only leadership can ensure that resources go to people, not projects, and that nurturing people's capabilities is given just as much weight as harvesting people's capabilities. Only leadership can ensure the smooth running of the organization for its central purpose, research.

Leadership is thus essential to a research organization. As articulated by Mervin Kelly, executive vice president (1943–1951) and president (1951–1959) of Bell Labs (Pierce, 1975, p. 209):

> When leadership is uninspired or inadequate, it is easy for research to drift away from the overall purpose of an organization. It is easy for the rest of the organization to disregard research. It is easy for systems engineers to become stale and to lose their feel for the actual state of research on one hand and the current realities of development, manufacture, and operation on the other. It is easy for a large staff organization concerned with buildings, facilities, shops, libraries, and even computer services to put organizational order and budgetary neatness ahead of the real needs and problems of scientists and engineers.

The common misperception that research is all about hiring outstanding researchers and letting them do what they want is far from the truth. Researchers of course contribute vitally to research excellence, but without research leaders, research organizations could not themselves remain vital. The decision to recruit and hire or promote a research leader is an extremely important organizational commitment, even more so than the decision to recruit and hire a researcher. Hiring outstanding researchers without outstanding research leadership already in place puts the cart before the horse. Underperforming research leadership, at any level of the organization, causes underperformance in entire swaths of the organization.

Moreover, outstanding research leaders are rare and their jobs, while different, are just as demanding as those of researchers themselves. Successful research leaders must have all the characteristics of outstanding researchers, particularly technical depth and breadth, otherwise their technical vision and judgment will not be sound. But they also need to be grounded in a second, orthogonal way: the human dimension that delights in and nurtures people's overall well-being and success. As articulated by VN years ago (Narayanamurti, 1987, p. 6):

> People don't just start innovating. People require an atmosphere where creativity can flourish, where ideas flow, and where personal development is encouraged. . . . I like to say that no one works for me, I work for my people. This is not meant to be just a cute line but is meant as a personal reminder that, even in a changing climate, today's managers must nurture—not manage—innovation.

Research leaders must embody multiple characteristics. They must be able to intuitively adjust depending on the person—they must have a "sense" of the person. They must provide compelling visions inspiring to people in the organization. They have a dual responsibility to lead people but also to serve them. They create environments that meet the needs of people—all their needs as complex human beings, of which their career is only one piece, albeit an important piece. They create secure bases for people—homes where people can go when they have nowhere else to go. They create communities that are empowered and that balance the bottom-up and top-down roles and responsibilities of researchers and research leadership. They create environments that are open and egalitarian: in which social and intellectual connections between people and topics are freely made and encouraged; in which inclusiveness, diversity, and "thinking different" are valued; in which administration is efficient but also subservient to research and not the other way around; and in which research leadership actively mentors with empathy and fosters linkages.

It is at the top levels of leadership—the director and overseers of the institution within which the research organization resides—that organizational purpose must be set. Only at this level can a "bigger than one-self" mission be articulated, defended, and given the longevity necessary for long-term success. And only at this level can the research organization as a whole be held accountable for its competitiveness in the larger national and global research landscape.

It is at the middle levels of leadership (deans of schools in universities and group managers or center directors in industrial or government laboratories), with guidance from the top, that organizational structure and resources must be set. Organizational structure and resources have many situational components that depend on particular knowledge domain (life sciences and medicine differ from physical sciences and engineering), on policies derived from the host organization (philanthropic research institutes differ from government laboratories or universities), and on geography and scale (organizations split across multiple sites may differ from those located at a single site). Leaders in this middle level, with their situational knowledge, must be engaged in architecting organizational structure and resources. These middle levels of leadership must also engage in critical strategic technical and personnel decisions, because they are close enough "to the ground" to form sound judgments of research directions and people, but one step removed with a broader view of how those directions and people fit with larger organizational strategy. Leaders at this level must also cultivate strong administrative skills. Indeed, if the reporting fanout becomes large enough, it may be necessary to bring on board executive-level administrators to support excellence in management of myriad organizational processes. In all cases, though, this support must be an executive, not an audit, function—executive administration at every level supports leadership; it is not a check on leadership. As articulated by a young postdoc about the extraordinary support given to research at Janelia Research Campus (Narayanamurti & Odumosu, 2016, p. 122):

The thing is that they set up this place so you really feel like everyone here, not just the postdocs or PIs [principal investigators], everyone here is here to make the research happen. So, they have facilities to make cell culture, to take care of animals. Even the guy who cleans the room or set up security procedures are here to make your job easier, to make the research job easier. For example, when we set up a new room for microscopes with lasers, the security folks came to our assistance to check out things for us and asked if we needed laser safety goggles or whatever equipment, and they purchased them for us. So, you feel like they are not trying to make their jobs easier, but they're here to make the research easy.

It is at the lowest levels of leadership, but also involving all levels, that comes the all-important day-to-day running of the organization. Research output is the combined product of researchers and research leadership, much like the overall competitiveness of a sports team is the combined product of athletes and coaches. In the remainder of this chapter, we discuss two aspects of this running of the organization: embracing a culture of holistic technoscientific exploration and nurturing people with care and accountability. Before we turn to these, we discuss a different aspect, one that more closely intersects organizational structure and resources: the building of critical mass.

As we observed earlier, research is often a collaborative activity and can be immensely more productive when individual researchers are surrounded by others with whom a breadth of ideas can be discussed with depth and mutual understanding—the interdisciplinary and transdisciplinary thinking that leads to next-adjacent possible combinations of ideas (Chapter 2). This we call *critical mass:* the nurturing of a critical number of researchers in a knowledge domain that has a just-right degree of breadth and depth.

On the one hand, an organization's research directions must be broad enough to accommodate the freedom to explore, learn, and opportunistically redirect to find the most impactful questions and answers.

On the other hand, an organization's research directions must not be too broad. With too much breadth, with continual unchecked expansion of research directions, three things happen. First, organizational knowledge in the research direction loses depth and cannot support a healthy intellectual competition and cross-checking of ideas and results. Second, research directions spanning too wide a cognitive distance can no longer be bridged effectively by researchers in the organization (Avina et al., 2018). As articulated by J. Rogers Hollingsworth in the case of the biomedical sciences (Hollingsworth, 2003, p. 221):

> Up to a certain point, more scientific diversity and communication increase the likelihood of breakthroughs in an organization, but when an organization has hyper-diversity with scientists focusing on so many problems, its scientific staff cannot communicate effectively to those in other fields. And, in such settings, there will be few major breakthroughs.

Third, when a research organization is thinly stretched, it become more difficult to maintain competitive advantage and benefit to the larger organization in which the research organization is embedded. Thinking strategically about the direction research should take—picking a niche that is unoccupied and that overlaps with the mission of the larger organization in which the research organization is embedded and pouring resources into it—is a characteristic of high-performing research organizations. It was natural, for example, for Bell Labs to choose communications as its broad research direction: doing so maximized the likelihood of breakthroughs important to AT&T's business mission and took advantage of synergies with AT&T's vast physical and intellectual infrastructure. It was natural for the National Institute of Standards and Measurements Physics Laboratory to choose measurement science, despite necessary and difficult decisions to terminate other areas (Ott, 2013). In the late 1970s, the University of California at Santa Barbara College of Engineering chose to focus on III-V compound semiconductors, since its small size precluded competition with larger engineering schools in mainstream semiconductors such as silicon (Narayanamurti & Odumosu, 2016). Finally, it made sense for Janelia Research Campus, founded in 2006, to focus on the intersection between the biological sciences and the physical sciences, computer sciences and engineering, because interdisciplinary collaborative excellence was its unique and designed-for institutional strength (Narayanamurti & Odumosu, 2016; Rubin, 2006).

Thus, an essential role of research leadership, at all levels but particularly at the lowest levels, is that of gardener: even as researchers extend their interests, which naturally leads to the blooming of hundreds of flowers, research leaders must weed out distractions so the research organization maintains its critical mass, focus, and strategic advantage relative to other competing research organizations. It is better to forgo a line of research than to enter into it without the resources needed to develop critical mass. Research leadership must fertilize the plants and judiciously kill the weeds—otherwise the flowers will die before they go to seed.

4.2 Embrace a Culture of Holistic Technoscientific Exploration

Just because organization, funding, and governance have been aligned with research—research has been funded to achieve metagoals not goals, research is insulated but not isolated from development, people not projects are block-funded, and active leadership is in place—does not mean that research will automatically be successful. Organization, funding, and governance alignment sets the backdrop against which research is executed but does not guarantee that the outcome of research will be surprise and creative destruction.

Exploring the technoscientific unknown also requires organizations to embrace a culture that holistically values *all* routes to technoscientific exploration: all the mechanisms of the technoscientific method, the finding of both questions *and* answers, and the seeking of surprise and the overturning of conventional wisdom. In principle, which route to emphasize at any time should be the route that situationally presents the most opportunity. In practice, however, some routes can become categorically undervalued and removed from the research organization's toolbox.

Here, we describe how various routes that are often undervalued must instead be embraced in a culture of holistic technoscientific exploration—and, as we do so, we place them in the context where they "fit" in our rethinking of the nature of research discussed in Chapters 1–3. In Chapter 1, we saw that $\dot{T}$ leads as much as follows $\dot{S}$, and thus research must not be relegated to $\dot{S}$ only. In Chapter 2, we saw that finding questions must balance finding answers. In Chapter 3, we saw that surprise to, must balance consolidation of, conventional wisdom—thus contrarians must balance conventionalists. Moreover, because surprise applies both to the particular outcome of research as well as to the long-term impact of research, research inspired by curiosity can be just as impactful as research inspired by practical application (Pasteur's quadrant) and both must be embraced.

$\dot{T}$ Leads as Much as Follows $\dot{S}$: Beyond Bush

Science and technology are intimately linked and symbiotic. A culture of holistic technoscientific exploration must thus embrace of the full technoscientific method. This principle seems obvious but, in fact, $\dot{S}$ and $\dot{T}$, and even subsets of $\dot{S}$ and $\dot{T}$, are often isolated from each other in blind adherence to the so-called linear model of innovation that sees science as the foundation on which technology is built. Although that model has been discredited repeatedly over the past decades (Godin, 2006), it still holds sway among many research policy makers and funders.

This tendency to favor science over technology at the front end of the research-and-development cycle gained momentum in the United States in the aftermath of World War II with Vannevar Bush's well-known "Endless Frontier" report. In his words (Bush, 1945, Cha. 1):

> Advances in science when put to practical use mean more jobs, higher wages, shorter hours, more abundant crops, more leisure for recreation, for study, for learning how to live without the deadening drudgery which has been the burden of the common man for ages past. Advances in science will also bring higher standards of living, will lead to the prevention or cure of diseases, will promote conservation of our limited national resources, and will assure means of defense against aggression. But to achieve these objectives—to secure a high level of employment, to maintain a position of world leadership—the flow of new scientific knowledge must be both continuous and substantial.

It is not much of a leap, then, to conflate science, assumed to be foundational, with research—where research is exploring the unknown and creating new knowledge paradigms and in that sense more truly foundational. The conflation was not universally made: it was not made by the great industrial research laboratories of the twentieth century nor by some of the research-funding organizations that drew leadership from them, including the Office of Naval Research under Emanuel Piore or the National Science Foundation under Erich Bloch. But the conflation is common and has even been codified by important research-funding organizations: the U.S. Department of Defense (their "6.1," "6.2," and "6.3" research and development [Sargent, 2018]), the National Aeronautics and Space Administration (their "technology readiness levels" [Mankins, 1995]), and the Organisation for Economic Co-operation and Development (their Frascati Manual [OECD, 2015]).

The origin of this conflation of research with science and of development with technology is not difficult to understand. The most singular advance of World War II, the atomic bomb, was a technological development whose origin lay in a scientific research discovery: the theory of special relativity. In this hugely important case, the research was of a scientific nature and the subsequent development was of a technological nature—in the atomic age, research *was* science, and development *was* technology. Moreover, this conflation benefited scientists, particularly academic scientists, who became a powerful interest group arguing for their monopoly role in the creation of the most foundational new knowledge (Pielke, 2012)—reminiscent of the early-twentieth-century snooty British-style elevation of gentlemen scientists, especially physicists, over engineers. And, as R&D statistics began to be collected, they were binned into categories commensurate with the linear model, thus entrenching the science-to-technology linear model even further as "social fact" (Godin, 2006).

In fact, every mechanism in the technoscientific method depicted in Figure 1-2 can be more or less exploratory, more or less research-like, in nature. As illustrated in Figure 4-4, $\hat{S}$ and $\hat{T}$ are independent of, and orthogonal to, the research and development continuum (Narayanamurti & Tsao, 2018; Tsao et al., 2008). On the one hand, research is not simply new science: it is paradigm creation whose outcomes cannot be scheduled or predicted in advance, a definition that cuts across science and technology. For example, the technological inventions of molecular beam epitaxy (MBE) (a way of synthesizing ultraclean materials layer-by-layer) and

	Disruptive or Radical Engineering	Revolutionary Science
Research Paradigm Creation and Surprise	*Invention* (A) Molecular Beam Epitaxy (B) Modulation Doping (D) High-Electron-Mobility Transistor	*Discovery* (E) Fractional Quantum Hall Effect (G) Possible New Particles…
Development Paradigm Extension and Consolidation	**"Standard" Engineering** *Improvement* (C) MBE of High-Electron-Mobility Heterostructures	**"Normal" Science** *Extension* (F) Deeper Investigations of Fractional Quantum Hall Effects
	Technology ($\dot{T}$)	Science ($\dot{S}$)

FIGURE 4-4. Four quadrants of knowledge evolution organized by their natures: research versus development and science versus technology.

modulation doping (a way of creating electronic carriers in a semiconductor without electron-scattering impurities) were just as much "research" as was the scientific discovery of the fractional quantum Hall effect and the ongoing quest for new particles with novel properties. On the other hand, development is not simply new technology: it is paradigm extension, an aspect of research that also cuts across science and technology. The subsequent deeper scientific investigations of the fractional quantum Hall effects were just as much "development" as was the ongoing technological improvement of MBE to produce ever-higher electronic mobility heterostructures. The observation of the not-yet-observed Higgs boson, because it was widely expected to exist, can be considered a paradigm-extending scientific outcome. Had it *not* been observed it would have overturned expectation and been a paradigm-creating scientific outcome.

Thus, we reject the conflation of $\dot{S}$ with research and $\dot{T}$ with development. In our view, science does not have a monopoly on research; nor does technology have a monopoly on development. Instead, both $\dot{S}$ and $\dot{T}$ encompass both research and development. $\dot{S}$ and $\dot{T}$ of a development character abounds and, as illustrated in Figure 4-4, these we might call "normal science" (Kuhn, 1962) and "standard engineering" (Arthur, 2009). $\dot{S}$ and $\dot{T}$ can also both have a research character and these we might call "revolutionary science" and "disruptive or radical engineering."

We call special attention to $\dot{T}$ with a research character—*engineering research*—as both extremely powerful and often underappreciated. Existing science can be a conservative force that denies the possibility of new technologies if they are inconsistent with current scientific understanding. But counter-to-current-science new technologies thrive and serve notice that nature, richer than we can imagine, is a fertile source of both new technology and new science. Prime examples come from the world of "materials discovery" (high T_c superconductors, efficient light-emitting GaN semiconductors, quantized-Hall-effect heterostructures, and 2D materials), all with properties not predicted by the science of the time. As articulated by Donald Stokes, acquiring and using expertise in technology often precedes acquiring and using expertise in science (Stokes, 1997, p. 95),

> Bush's claim that "a nation which depends upon others for its new basic scientific knowledge will be slow in its industrial progress and weak in its competitive position in world trade" . . . was disputed by America's own experience in the earlier decades of the century when it borrowed the European science and technology it needed to become the world's leader in industrial technology while lagging in basic science. Plainly, Japan's success owes more to acquiring and improving upon the world's technology, including a good deal that is science based, than to building a basic-science dynamo to power its industrial progress from within.

Engineering research grounds scientific research in the real world and in real problems; it is a key conduit through which nature surprises us and forces us to be creative; and it is a key cross-check on science.

Thus, nurturing research must nurture $\dot{T}$ just as much as it nurtures $\dot{S}$. Engineering research, $\dot{T}$ with a research flavor, is so often the initial seed for the other kinds of knowledge advance: $\dot{T}$ with a development flavor, $\dot{S}$ with a research flavor, and $\dot{S}$ with a development flavor. This is illustrated by the sequence of knowledge advances surrounding the fractional quantum Hall effect listed here chronologically (A) to (G) and shown grouped in the Figure 4-4 quadrants:

A. Molecular beam epitaxy (MBE) was invented in the late 1960s (Cho, 1971).
B. Modulation doping was invented in 1977 (Dingle et al., 1978).
C. The combination of these two inventions led to MBE of increasingly tailored (and high electron mobility) heterostructures (Störmer et al., 1981).

D. The high-electron-mobility transistor was invented in 1979 (Dingle et al., 1979; Mimura et al., 1980).

E. The fractional quantum Hall effect was observed in 1981 (Tsui et al., 1982).

F. Deeper investigations of such effects have continued subsequently (Stormer, 1999).

G. The quest continues for possible new charged particles in reduced dimensions obeying fractional statistics (Bartolomei et al., 2020).

Two seminal engineering, not scientific, research advances, (A) molecular beam epitaxy and (B) modulation doping, were the initial seeds for the rest.

Type of Inspiration Is Not a Good Leading Indicator of Type of Impact: Beyond Pasteur's Quadrant

A distinction commonly made about both $\dot{S}$ and $\dot{T}$ is how they were inspired: was the doing of $\dot{S}$ and $\dot{T}$ inspired by curiosity or practical application? We illustrate that distinction using typology similar to the "Pasteur's quadrant" typology popularized by Donald Stokes (Stokes, 2011) but expanded to include the research-versus-development distinction discussed just above (Figure 4-5). The four quadrants at the front are the "research and surprise" quadrants; the four quadrants at the rear are the "development and consolidation" quadrants.

For both sets of quadrants, the bottom axis organizes knowledge advance according to how it was initially inspired: by practical applications, such as eating, versus by human curiosity, such as discovering the origin of the universe or inventing a laser just to see what it can do. In Stokes's language, these are "considerations of use." Also for both sets of quadrants, the left axis organizes knowledge advance according to whether it is of a scientific ($\dot{S}$) versus technological ($\dot{T}$) nature—whether it lies in the "scientific method" or "engineering method" half of the overarching technoscientific method. In Stokes's language, this is whether the knowledge advance is associated with a "quest for fundamental understanding," which he equates with science, or nonfundamental understanding, which he equates with technology.

The four research quadrants are depicted at the front of Figure 4-5. Three of the quadrants are named after Stokes's original exemplars but with our descriptors: Bohr's quadrant, curiosity-inspired scientific research; Pasteur's quadrant, application-inspired scientific research; and Edison's quadrant, application-inspired engineering research. The fourth quadrant, in white, was unnamed by Stokes. Because of the prejudice against technology in the linear model, this quadrant is sometimes called the "stamp-collecting" quadrant, as if there were nothing significant to be gained from curiosity-inspired exploration of technologies other than what can be inspired by known practical applications. Instead, we name this quadrant Townes's quadrant, curiosity-inspired engineering research. As exemplified by Charles Townes and the engineering invention of the maser / laser (Chapter 1), significant technology advance can be inspired by curiosity just as much as it can by practical application. Building a device that emits light more coherently, more uni-directionally, and with narrower spectral line widths than required by any known practical application—this satisfied human curiosity just as much as did Bohr's scientific understanding of the internal structure of the atom and was just as surprising and profound a knowledge advance. But it did not satisfy any then-known practical application: the usefulness of both the maser and the laser were doubted by Charles Townes's own research leadership at Bell Labs, which partially motivated him to move from Bell Labs to Columbia University to continue that research. Even after the laser was invented, it was, in its early years, jokingly called "a solution looking for a problem" (Hecht, 2010, p. F104), a phrase still used in derision to refer to answers seeking questions to answer.

To these four research quadrants, we add the analogous four development quadrants to the rear in Figure 4-5. The Whafnium quadrant, curiosity-inspired scientific development, is the curiosity-inspired extension of scientific knowledge known about one material (the imaginary "whoofnium") to a closely related material (the equally imaginary "whafnium") (Goudsmit, 1972). The Si-Properties quadrant, application-inspired scientific development, is the extension of scientific knowledge specifically for the purpose of improving a practical application, as epitomized by the scientific study of the electronic, thermal, mechanical, and chemical properties of Si for the purpose of optimizing the performance of Si-based electronics. (It is sometimes said that Si has been the most

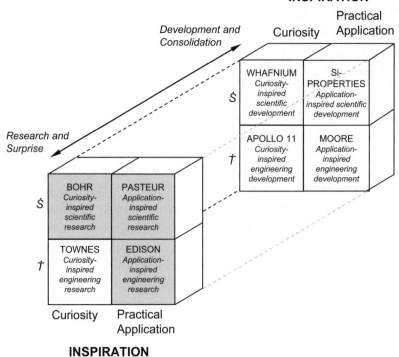

FIGURE 4-5. The fourfold "Pasteur's Quadrant" typology of Donald Stokes, doubled to include both research and development.

studied of all inorganic elements.) Moore's quadrant, application-inspired engineering development, is the engineering improvement of a technology for the purpose of improving its use in an application—much like Moore's law describes the continuing improvement of digital integrated circuits. The Apollo 11 quadrant, curiosity-motivated technological development, is the extension of known engineering in service of curiosity—like the monumental US engineering development of the technology for human space flight to the moon and back, not for its practical utility but for satisfying overarching human curiosity.

Limitations of Quadrant Terminology: Type of Impact Does Not Follow from Type of Inspiration

The two sets of quadrants just enumerated are logical extensions of Stokes's framing, but we have been careful not to use his terminology which, while popular, is limiting. Stokes describes Bohr's quadrant as "pure basic research" and Pasteur's quadrant as "use-inspired basic research." Stokes describes Edison's quadrant as "pure applied research," though a more internally consistent descriptor would be "use-inspired applied research." Stokes left the fourth quadrant unlabeled, but an internally consistent descriptor would be "pure applied research." Using Stokes's terminology, the two terms "basic" and "pure" have come to mean the following: "basic research" is research that is of a scientific flavor hence describes the top two quadrants; "applied research" is research that is of a technological flavor hence describes the bottom two quadrants. "Pure research" is research that is inspired by curiosity rather than practical application hence describes the left two quadrants; "use-inspired research" is research that is inspired by practical application rather than curiosity hence describes the right two quadrants.

How is this terminology limiting? It is limiting in two ways.

First, the adjective "basic" carries with it the sense of "foundational." So when it is also used to mean research that is of a scientific flavor, it conflates scientific research with more foundational research. Reinforced by the "linear" model of the evolution of technoscience, we then easily jump to an incorrect conclusion: that scientific research is the more foundational research on which technological (engineering) research builds. In fact, the degree of paradigm creation and surprise, hence the degree to which the knowledge advance is truly foundational, does not depend on its scientific or technological flavor. It depends solely on its degree of surprise—the degree to which it contradicts conventional wisdom and forces the researcher and the scientific community to change course. Edison and Townes upended conventional wisdom just as much as did Bohr, and their knowledge advances were just as foundational.

The second way in which it is limiting is in the term "pure," which connotes research unsullied by practical considerations and therefore unable to impact practical application. Though this conclusion in some cases may be true for development, it is by no means true for research. Research is characterized by the metagoal of surprise. Surprise means both that the particular knowledge advance cannot be anticipated and, just as importantly, that the subsequent downstream impact of that knowledge advance cannot be anticipated. Research inspired by curiosity can lead to just as significant an impact on practical application as that inspired by practical application, as illustrated by Michael Faraday's famous response to the question asked by William Gladstone, then British chancellor of the exchequer, about the then-doubtful practical value of electricity. In his own defense Faraday said,

> One day, sir, you may tax it.

In other words, whether research is inspired by curiosity or by practical application is not a good leading indicator of the ultimate impact of the research. Moreover, research inspired by one practical application can lead to a significant impact on an unintended other practical application—as in the transistor revolutionizing computation but not, at least at first, revolutionizing communications. As illustrated in Figure 4-2, due to the mechanisms of generalization and exaptation, impact outside the originally intended knowledge domain is the rule, not the exception. Moreover, such spillover impact is not a matter for despair; it is simply a matter of relinquishing control where control is impossible. As articulated by computer scientists Ken Stanley and Joel Lehman (Stanley & Lehman, 2015, Loc. 1450):

> The situation may sound hopeless, but the conclusion is both subtler and more profound than it seems: We can reliably find something amazing. We just can't say what that something is! The insight is that great discoveries are possible if they're left undefined.

Limitations of Quadrant Terminology: Opportunistic Shifting Across Quadrants Is the Rule, Not the Exception

The source of inspiration is not only not a good indicator of the ultimate impact of the research but is also a moving target. Surprise is the metagoal of research, but where the most opportunity for surprise is to be found is constantly shifting as knowledge spaces are explored. Research that begins in one quadrant can easily and profitably shift into other quadrants. Quantum mechanics, begun in Bohr's quadrant, shifted into Pasteur's and Edison's as it inspired important practical applications such as semiconductor devices. The laser, begun in Townes's quadrant, shifted into Bohr's and Edison's as it inspired both the observation of new light-matter interactions and practical applications such as optical fiber communications. The fractional quantum Hall effect, which had its origins in Pasteur's quadrant, shifted into Bohr's quadrant due to its importance as a signature of strongly correlated electron phenomena.

Research is about what cannot be anticipated, and part of what cannot be anticipated is which quadrant will be most profitable to explore at any given time. As new knowledge is gained, the opportunistic switching between quadrants must be supported. The absence of such opportunistic switching is what limits the potential impact of Google X, Google's research arm. Because their filter in deciding what research to pursue is "10x improvement in societal impact" (Thompson, 2017), they are automatically dismissing Bohr's and Townes's quadrants. It is also what is missing from the so-called ABC (applied and basic combined) principle of research (Shneiderman, 2016), which emphasizes Pasteur's quadrant. Advances in technoscience can sometimes be inspired by practical application, and inspiration from practical application can sometimes lead to advances in technoscience—but not always, so it should not be forced. For maximum research impact, one must expect the unexpected, including shifts from quadrant to quadrant.

Note also that opportunistic shifting across quadrants can even happen across research and development. One can begin in a development quadrant, then notice an anomaly. With a development mindset, the anomaly might be ignored or attributed to experimental error rather than to a breakdown in one's understanding of what to expect. With a research mindset, however, the anomaly would be looked at as an opportunity to find unusual behavior that contradicts conventional wisdom and may lead to an unexpected discovery.

Embrace Finding Answers *and* Questions: Beyond Heilmeier

As presented in Chapter 2, using an evolutionary biology analogy, answer-finding is adaptive, downward-looking, and reductionist, while question-finding is exaptive, upward-looking, and integrative. An answer-finding mentality focuses on known questions but unknown answers and requires deep understanding of the solution space of a given discipline. A question-finding mentality focuses on known answers but unknown questions and requires deep understanding of the problem space encountered by a given user. A culture of holistic technoscientific exploration encompasses both question- *and* answer-finding.

Though both question-finding and answer-finding are critical to research, typically much more emphasis is placed on finding answers than on finding questions, as exemplified by the "Heilmeier Catechism" of the Defense Advanced Research Projects Agency. Proposal writers are asked to respond to the following (Cheung & Howell, 2014, p. 12):

> What are you trying to do? Articulate your objectives using absolutely no jargon.
>
> How is it done today, and what are the limits of current practice?
>
> What is new in your approach and why do you think it will succeed?
>
> Who cares? If you are successful, what difference will it make?
>
> What are the risks and the payoffs?
>
> How much will it cost?
>
> How long will it take?
>
> What are the mid-term and final "exams" to check the success?

A research proposal that satisfies the Heilmeier Catechism poses a question and proposes to answer it. Ideally, such a proposal is predictable and schedulable—one knows what tasks and milestones one is trying to accomplish and lays out a plan to accomplish them.

Such a proposal thus misses half of the question-and-answer finding coevolutionary dance. Coevolution may start with a particular question that guides the research, but in the process of trying to answer it, a way to answer a different and perhaps even more important question might be found, and researchers should flexibly change directions to run with the new question. Indeed, George Heilmeier himself was known for this kind of visionary question-finding, even if it is paradoxically not reflected in his own catechism. As articulated by his colleague Stu Personick (Cheung & Howell, 2014, p. 13):

> George had the ability to recognize the prospective, far reaching implications of a new discovery such as the underlying ferroelectric effects in liquid crystals, to create solutions for important unmet market needs, and to follow through on the applied

research and development needed to turn a new discovery into high impact, marketable products.

Just as Heilmeier knew the importance of question-finding, so did Pasteur. He was famously able to take chance events and observations, realize that these were relevant to a different question than the one he had set out to answer, and flexibly switch his line of research to the new question. A similar question-and-answer coevolution was evident in the modulation doping and fractional quantum Hall effect work discussed earlier. Flexibility in accommodating new findings, the degree to which the work is not confined to a narrow question-and-answer space, correlates strongly with whether knowledge production is more research- or development-like.

The Scientific Method Is More Than Hypothesis-Testing: Beyond Popper

The prevailing emphasis on finding answers rather than finding questions is also illustrated by the increasingly popular equating of the scientific method with hypothesis-testing (a kind of answer finding). But hypothesis-testing, though fast becoming a necessity in scientific research proposals, is only one piece of the scientific method. The scientific method, as outlined in Chapter 1, has *three* mechanisms.

The first mechanism of the scientific method is fact-finding: the use of technology or its antecedent, human senses, to elicit observations about phenomena. Fact-finding includes hypothesis-testing: the deliberate seeking of phenomena predicted by theory; hence if the phenomena is found, the theory is more plausible; if not found, the theory is less plausible. But it also includes the seeking of phenomena in an open-ended way, where the hope is to find the unexpected. To do so, typically one would propose to expose a technology to an environment or to conditions that the technology to which it had not previously been exposed. The more "different" or "extreme" the environment (high pressures, low temperatures, high energies), the more likely new phenomena will be observed. However, while one can create conditions that would make seeing the unexpected more likely, one cannot plan to see the unexpected, and to the extent that most proposals require plans and milestones, fact-finding of the unexpected is nontrivial to center a research proposal on.

The second mechanism of the scientific method is explanation-finding, the explaining of existing stylized facts. Explanation-finding is often underappreciated and undersupported compared to fact-finding and, especially, to hypothesis-testing. A scientific research proposal must typically be observational in nature: either proposing to find new facts from observations or already containing a hypothesis to test through observation. If it only contains preexisting observations one wants to try to make sense of, the proposal will likely not get funded because of the huge uncertainty associated with how explanations might be found. As articulated by Richard Feynman (Feynman, 2015, p. 291):

> When you're thinking about something that you don't under-
> stand, you have a terrible, uncomfortable feeling called confu-
> sion. It's a very difficult and unhappy business. And so most of
> the time you're rather unhappy, actually, with this confusion.
> You can't penetrate this thing. Now, is the confusion because
> we're all some kind of apes that are kind of stupid working
> against this, trying to figure out [how] to put the two sticks to-
> gether to reach the banana and we can't quite make it, the idea?
> And I get this feeling all the time that I'm an ape trying to put
> two sticks together. So I always feel stupid. Once in a while,
> though, the sticks go together on me and I reach the banana.

Because explanation-finding is so important, however, scientists will do it anyway, but without formal support. As articulated by Herbert Simon (Simon, 1977, p. xvi):

> The philosophy of science has for many years taken as its central
> preoccupation how scientific theories are tested and verified,
> and how choices are made among competing theories. How the-
> ories are discovered in the first place has generally been much
> neglected, and it is sometimes even denied that the latter topic
> belongs at all to the philosophy of science. . . . This emphasis
> upon verification rather than discovery seems to me a distortion
> of the actual emphases in the practice of science. . . . The history
> and philosophy of science have been excessively fascinated with
> the drama of competition between theories: the wave theory of
> light versus the particle theory, classical mechanics versus

relativity theory, phlogiston versus oxygen, and so on. Such competition occurs only occasionally. Much more often, scientists are faced with a set of phenomena and no theory that explains them in even a minimally acceptable way. In this more typical situation, the scientific task is not to verify or falsify theories, or to choose between alternative theories, but to discover candidate theories that might help explain the facts.

In contemporary physics, prominent examples are the fractional quantum Hall effect, superfluidity in liquid helium-3, and unconventional (high-temperature) superconductivity—all observations that predated explanation.

The third mechanism of the scientific method is generalizing: the extending of theory, originally intended to explain one set of facts, to predict possible facts it was not originally intended to explain. When those possible facts have already been observed, then the theory is immediately given more credibility. When those possible facts have not yet been observed, then one has found a hypothesis—a potential scientific fact that, if it could be observed, would give the theory more credibility. The potential scientific fact that starlight would undergo a particular angular deflection by gravity as it passes near the sun was a hypothesis that, when shown correct by Arthur Eddington in 1919, gave Einstein's theory of general relativity credibility. As simple as many hypotheses seem after they have been found, however, finding them is nontrivial. Like all generalizations, the finding of new questions is difficult to anticipate and plan ahead of time, so hypothesis-finding is also difficult to support via planned projects.

All this is not to argue against the importance of hypothesis-testing. Hypothesis-testing is of obvious importance to the scientific method. Two classic dictums show how the scientific method differs from other methods of acquiring knowledge. The Baconian dictum *"nullius in verba"* (don't trust words) is the motto of the Royal Society, the world's oldest independent scientific academy. The Popperian dictum asserts that the ability of a theory to be falsified through hypothesis-testing is the only true test of whether a theory is scientific or not. Without hypothesis-testing, we have superstition and "cargo cult" science (Feynman, 1974). But just because hypothesis-testing is important does not mean it is all-important or that it fully defines the scientific method. The scientific method comprises all three of the mechanisms discussed above, not just that piece of the fact-finding mechanism that represents hypothesis-testing.

Embrace Informed Contrariness: Beyond Peer Review

Another hallmark of a culture of holistic technoscientific exploration is the embracing of informed contrariness. This is important because, as discussed in Chapter 3, surprise and creative destruction at their most extreme are the finding of implausible utility—useful advances in knowledge that are surprising.

How might one find the surprising?

One might go with conventional wisdom in deciding where to look, but find something that conventional wisdom did not expect. Then, as implausible as the apparent finding might appear, take it seriously enough to nurture it into an invention or discovery. In the example of Figure 4-4, it made "sense" to look for new electronic phenomena when heterostructures with unprecedented electron mobility became available through advances in materials synthesis technology, but the observations indicating fractional rather than integer electronic charge were totally unexpected. They could have been dismissed as observational error or noise but weren't, and the result was the 1998 Nobel Prize in Physics, awarded to Horst Stormer, Dan Tsui, and Robert Laughlin (Stormer, 1999).

Or one could go against conventional wisdom in even deciding where to look and, on top of that, find something that conventional wisdom didn't expect. It made no "sense" for Galileo to train a telescope on the heavens given, as mentioned earlier, Aristotle's pronouncement that celestial objects were perfect and immutable spheres. Yet, in doing so, Galileo found the moons of Jupiter—something totally unexpected by conventional wisdom. It made no "sense" for Torricelli to weigh the air much less to find that it in fact had weight.

Both routes share a common willingness to take seriously observations or ideas contrary to conventional wisdom. Because conventional wisdom is limited, it is precisely when a researcher's observation or idea goes beyond (and is even implausible to) conventional wisdom that the potential for implausible utility is greatest. Implausible utility requires researchers to suspend belief in a portion of conventional wisdom and to take seriously contrarian alternatives: that the earth's continents drift, that the speed of light is constant regardless of reference frame, that the iPhone can fulfill human-desired function, or that audio signals can be transmitted over electrical wire.

Unlike plausible utility, implausible utility is a lonely path. As indicated by the right side of Figure 4-6, any observation or idea whose utility could have been predicted in advance by conventional wisdom ($u_{ConvWis}$) is by definition *not* going to lead to implausible utility. It is "dead on arrival" with respect to successful research outcome precisely because it was *plausible*. Only an observation or idea on the left side of the diagram, one that is *contrarian*, has a chance at implausible utility. Contrariness, however, is difficult for most researchers and research organizations to accept. Contrariness requires the ability to withstand criticism from, even the dismissiveness of, peers. Indeed, peers are often right to be dismissive: history is littered with observations and ideas dismissed at the time and ultimately shown *not* to have utility. The initial dismissal of reports of cold fusion in 1989 turned out to be justified. Nonetheless, those contrarian observations and ideas that do succeed can be transformative: history is also littered with observations and ideas that were dismissed at the time but ultimately proved to be extremely fruitful.

Thus, researchers and research organizations who aim to "change the way people think or do" must have the freedom, not only to be contrarian, but also to be wrong. In the context of his time, Edison in 1899 wrongly thought that electric cars would be superior to gasoline- or steam-powered vehicles and in 1902 wrongly thought that batteries would be invented that would enable driving of one hundred miles or more before recharging. But because he had the freedom to be contrarian, to be wrong, Edison also had the freedom to try and to succeed in making other world-changing inventions, including the electric light bulb and the electricity generation and distribution systems discussed in Chapter 3. Contrariness sometimes leads to failure, but from failure comes learning, and from learning very often comes implausible utility, the useful and surprising.

Contrariness is not the only thing required of researchers to achieve implausible utility, however. The second thing that is required is *informedness.* Conventional wisdom and existing paradigms "work"—that is why we adopt them in the first place and that is why we resist so strongly their overthrow. If a researcher is going to take seriously observations and ideas that go against conventional wisdom, the researcher had better have good reasons for doing so—and the discipline to develop those good reasons. These reasons we call informedness—"inside" knowledge or capabilities the researcher possess that the researcher's peers don't yet have. This inside knowledge makes the researcher think the researcher is right and conventional wisdom is wrong. The researcher is an "informed contrarian," going against conventional wisdom but in an informed way to reduce the tremendous risk associated with going against that very wisdom.

Like a financial arbitrageur who uses greater informedness about the true value of an asset to buy those assets currently undervalued by conventional wisdom, informed contrarian researchers are research arbitrageurs who use their greater informedness about the value of a research observation or idea to take seriously those ideas currently undervalued by conventional wisdom. To maximize research arbitrage, we want to be as far into the upper-left corner of Figure 4-6 as we can be. Researchers seek observations or ideas whose value is high according to their informedness (u_{Inf}) but low according to conventional wisdom (u_{ConvWis}). A close analogy can be made to venture capitalists choosing which start-ups to back. As well-known venture capitalist Peter Thiel puts it to entrepreneurs he might invest in, "Tell me something that's true that almost nobody agrees with" (Hof, 2014, p. 1). To be playful with Pasteur's famous saying, one might say "contrariness favors the informed mind."

So the essence of what we are looking for in a researcher is *informed* contrariness. Informed contrariness does not guarantee success; the researcher might still fail. But uninformed contrariness is worse. It is equivalent to being in the bottom left quadrant of Figure 4-6, where one hopes without any informed confidence but perhaps out of desperation that one's contrariness will pay off—like a "Hail Mary" pass in American football.

At the researcher level, informed contrariness requires balancing temperamental extremes: the predisposition to "think different" yet to rigorously test that "different thinking"; a passion for the unknown yet

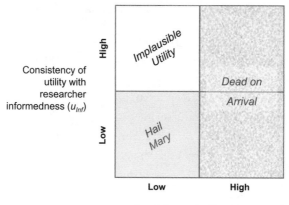

FIGURE 4-6. The route to implausible utility.

respect for the known; comfort with necessary risk and uncertainty but discomfort with unnecessary risk and uncertainty; an obsession with and curiosity for what one doesn't know *and* a healthy respect for what one does (or should) know; the self-confidence to trust one's own views yet the self-criticalness also to distrust one's own views; and the personal ambition, the so-called fire in the belly, necessary to question and overturn conventional wisdom but the honesty and integrity to respect truth and change one's own view when warranted.

At the research organization level, informed contrariness requires a similar balancing of extremes. On the one hand, the research organization must encourage new and contrarian ideas even if these ideas are contrary to the organization's own beliefs, and it must protect the quirky and eccentric researchers from whom such ideas often emerge. On the other hand, the research organization must encourage criticism of those ideas to weed out ideas that truly make no sense, or at least no sense to pursue at that moment. Especially when one is being contrary, one needs to be informed, and being informed requires benefiting from others' knowledge and critiques. *En route*, it creates a certain amount of stress, but one would no more expect elite researchers to achieve maximum performance without stress than one would expect the same of elite athletes. Indeed, the opposite of this culture of truth, a culture of consensus, is a prescription for social harmony, conflict avoidance, group-think, and lower stress rather than for informedness in the face of contrariness (Avina et al., 2018).

4.3 Nurture People with Care and Accountability

As just discussed, aligning organization, funding, and governance for research, and embracing a culture of holistic technoscientific exploration—both are important to the success of any research enterprise. Ultimately, though, research is a human endeavor, and to succeed at this endeavor at a high level also requires nurturing people at a high level. We draw attention here to three key aspects of that nurturing: (1) recruiting, hiring, and mentoring; (2) holding research accountable; and (3) the practical instantiation of accountability, performance review. Together, they represent a combination of genuine personal care for the human being and high standards of professional accountability for research results. There is the empathy and care associated with recruiting, hiring, and mentoring; with joint nurturing of researchers and research directions; and with rewarding for more successful research outcome. There is also the selectivity and accountability associated with *not* hiring, with weeding out research directions no longer at the research frontier or that have morphed too far afield from the overarching mission of the organization, and with holding researchers accountable to high standards of research outcome.

Recruiting, Hiring, and Mentoring

At the front end of nurturing people for research is recruiting, hiring, and mentoring. The commitment to recruit, hire, and mentor a researcher—to give a researcher long-term support and freedom—is a major one. Every high-performing research organization must pay extreme attention to recruiting, hiring, and mentoring.

Recruiting and Hiring: Technical Quality, Passion, and Diversity

Top-notch researchers are rare and precious, and every effort must be made to get to know them as both human beings *and* intellects as they are being recruited. Relationship-based processes are critical: personal visits to university campuses to chat informally with students and faculty, taking personal interest in noteworthy talks given at technical conferences or in noteworthy papers published in technical journals, and deep intellectual probing of candidates before and during on-site visits or interviews. Involvement of research leadership at all levels of the organization is critical: the fit might be wrong for any number of reasons, and among the most important qualities of leadership is the ability to assess fit. As recounted by Art Gossard (who made seminal contributions to the fractional quantum Hall effect work discussed earlier in this chapter) about his experience being recruited at Bell Labs in the 1960s (Narayanamurti & Odumosu, 2016, pp. 90–91):

> So, for me, I went through graduate school in Berkeley, and there I experienced the Bell Labs recruiting, which was very special, and the Bell Labs recruiters visited Berkeley every year and they followed up the progress of the students in Berkeley as they did at many, many other universities. So that was a very effective recruitment tool for the labs. It was also good for me. My recruiter was Bill Boyle, who at that point was a recruiter and department head and subsequently won the Nobel Prize for charge-coupled devices. So, he lined up my visits at Bell Lab and that consisted of two days of visiting both the research area where I was interested in being a postdoc and visiting many, many other areas.

This kind of deep relationship-based recruiting must be maintained even when "post and bid" processes are used to widen recruitment scope and enable a more diverse recruitment pool than is possible solely with personal and professional contacts. Getting to know the human being behind the job application, really understanding their research interests and abilities, is critical. Initial impressions are not necessarily good predictors of future performance. Someone who comes across initially as dense and slow might actually be deep and insightful. One can only know this if one takes the time to know the candidate as a human being, including getting to know the references that the candidate gives and getting a feel for how much to trust particular references.

Exactly what to recruit and hire for is, of course, tricky, and it is fair to say that no one has yet found the exact formula. Elon Musk's "clear evidence of exceptional ability" begs the question of "what abilities?" A few dimensions of these abilities, though, do seem clear.

A first dimension is technical depth, a depth of knowledge in one or more technical areas. This dimension is somewhat easier to evaluate, as it maps to academic performance—to an understanding of current paradigms in a topic area. Note, though, that one should be somewhat neutral to the exact technical area of depth, and indeed not to be neutral is one of the easiest mistakes that research organizations make. The organization "needs" a person who knows X because X appears to be essential for exploring a particular research direction. But research outcome cannot be anticipated, and as research progress is made, it may soon be Y or Z that is more important to know. The importance of knowing X deeply is, thus, not because X is *itself* important, but because knowing X deeply is a leading indicator of an ability to learn Y and Z deeply as research takes its twists and turns. One should not underestimate the ability of great researchers to move from area to area as their interests and necessity take them.

A second dimension is technical breadth, along with an innate curiosity for exploring beyond one's specialty. This is, of course, more difficult to assess when recruiting young staff, who have generally had a chance to prove themselves within a single discipline but not yet across disciplines. Typically, one must look for intangibles—the surprise and sparkle in their eye when a new analogy is mentioned that perhaps they hadn't thought of before, their willingness to explore new ideas in real time, their excitement (even if combined with some fear) when the prospect is raised of doing something totally different. At Bell Labs, the famous "empty lab" question (What would you do with an empty lab and no constraints on budget?) was meant in part to assess curiosity for exploring the unknown.

A third dimension is passion, which is important in every creative walk of life and no less so in research. A passion for exploring the unknown. A passion for making an impact. A passion that drives a person to learn, to achieve, to want to be surrounded by those who themselves love to learn and achieve and to work harder when given more freedom and less structure. This dimension is typically underrated but is extremely important. A hunger to achieve is key to actually achieving.

A fourth dimension is diversity, a dimension that has to do with success of the organization as a whole, not just success of the individual. By diversity, we mean diversity in its broadest sense: in temperament, in area of technical expertise, and in experiential background that includes gender and cultural origin. Such diversity is key to the collaborative and interdisciplinary finding of new and surprising questions and answers, and it will be key to the creative addressing of major human and societal challenges of the future. Great research leaders must guard against "old boys" hiring for uniformity and instead build an organization with a diversity of human and technical perspectives that stimulate divergent thinking and creativity—and that often stem from women and other under-represented groups, and from brilliant eccentrics and loners. Great research leaders must manage such diversity with empathy and care, however, as it can easily fall apart—not all idiosyncratic individuals have social graces and teaming skills. The heterogeneity of the human and intellectual condition must be recognized and valued.

Of these four dimensions, technical depth and breadth seem easier, and passion and diversity seem more difficult, to evaluate. But all dimensions are nontrivial to evaluate. Quantitative metrics can and should be collected and considered, but, ultimately, whom to recruit and hire is (and must be) a human judgment, even as that judgment should be subject to higher-level scrutiny. It cannot be left to a formula. Following researchers early in their careers, even while they are still students; assessing whose recommendations and references to trust; getting a sense for whose motivations, interests, and capabilities are most compatible with the organization's broad research directions and with the potential for research excellence—these are refined and multi-dimensional judgments that only research leaders can make. Research leaders will make mistakes, of course; human judgment is fallible. But research leaders must nonetheless exercise their human judgment, and those who are more adept (even if no one is able to articulate why they are more adept), should be given more responsibility to exercise that judgment.

Deciding whom to recruit and hire is one thing; attracting the person to actually join the organization is another. Candidates are human with varied preferences for geographic location, for different overarching organizational missions, and for many factors not under the control of the organization. When these factors aren't right for a person, one would be doing them a disservice to overly encourage them to join. It is a matchmaking process, with both sides evaluating both sides. One factor, though, is universal: the best researchers are attracted to work with the best researchers and at the best research organizations, so the research quality and the reputations of researchers and research leadership are critical.

Mentoring

Once hired, a researcher embarks on exceedingly difficult tasks: to explore the unknown, to choose promising areas to dive into, to choose collaborators, and, ultimately, to overturn conventional wisdom in some way. As they embark on these tasks, researchers require human nurturing and mentoring by astute and caring research leaders. There is no substitute for research leadership that actively mentors; understands strengths, weaknesses, and unique passions; helps in an inclusive manner to make connections with potential collaborators and emerging research directions; encourages going after the new and unknown; removes impediments and levels the playing field, especially for new employees, women, and minorities; cheerleads research and human successes; and, above all else, pushes for the exploration of the unknown. As has been said about the mentoring given by Mervin Kelly, president of Bell Labs from 1951 to 1959 (Gertner, 2013, p. 109):

> Pierce later remarked that one thing about Kelly impressed him above all else: It had to do with how his former boss would advise members of Bell Labs' technical staff when they were asked to work on something new. Whether it was a radar technology for the military or solid-state research for the phone company, Kelly did not want to begin a project by focusing on what was known. He would want to begin by focusing on what was not known. As Pierce explained, the approach was both difficult and counterintuitive. It was more common practice, at least in the military, to proceed with what technology would allow and fill in the gaps afterward. Kelly's tack was akin to saying: Locate the missing puzzle piece first. Then do the puzzle.

Research leadership is not just tasked with nurturing researchers, as complex as that task is. Research leadership is responsible for joint nurturing of both researchers *and* research directions, a task that is far more complex, as sometimes these are in harmony, and sometimes not.

From the perspective of the research directions, leaders must ask, "What are the most interesting directions on the horizon, ripe for advance?" Sometimes the new directions are variants of existing directions, so with minimal reshaping they might be amenable to attack by the talents, expertise, interests, and career trajectories of existing researchers in the research organization. But sometimes the new directions do not share commonalities with existing directions but are nonetheless extremely important to the organization. Then, difficult choices must be made. Are new researchers, with the appropriate talents, expertise, interests, and career trajectories recruited and hired? Are existing researchers encouraged to move into the new directions from their current work?

From the perspective of the researchers, leaders must ask, "What *are* the talents, expertise, interests, and career trajectories of our researchers? How might these be applied to match the organization's research directions as well as enable the researchers to learn and be challenged in directions consistent with their personal and professional goals?" In many cases, researchers will self-assemble into collaborations and research directions that they are optimally challenged by and can productively pursue. Particularly when researchers have a history of wise decisions on the research directions they have pursued, much freedom should be given. If a researcher is on the track of something they view as important, even if the impact of the research direction appears less than plausible, then they must be free to make that choice. It is a common refrain from society's most celebrated researchers that their best research occurred despite their peers' belief in its implausibility. Research leaders can and should inspire new directions, and most researchers will be open-minded to consider those directions, but for the best researchers it would not be smart in general to force the choice.

In some cases, though, self-assembly will be suboptimal. Researchers are not all-knowing—particularly researchers outside of the social mainstream or at the beginning of their careers. Sometimes they are unaware of potential collaborations that would unlock new perspectives and approaches. Sometimes they get trapped into working in an area that has reached a level of maturity where new results are not likely— resting on their reputation and / or fearful of the uncertainty and challenge of moving into new areas. Encouraging, sometimes insisting, that such researchers find new collaborators or move on to new research directions—these are essential responsibilities of research leadership. As articulated by Ralph Bown, vice president of research at Bell Labs from 1951 to 1955 (Bown, 1953, p.4):

> The hardest job that researchers, and in particular research supervisors and directors, have is to let go of new things just when

they get exciting in an applicational way and turn back to their job of groping for the next good idea. If they fail to exercise this self-control and wisdom and let themselves be drawn into competition with the development engineers, their end as a vital research organism is in sight. Dropping the old to grasp the new is the hardest and most vital decision that research groups are called on to make.

Hold Research Accountable

It is a common misperception that research is a playground, a place for play, not work, a place for fun and games without judgment. There are, of course, elements of that analogy that are correct: the emphasis on curiosity, learning, and exploring. But researchers in such a playground would quickly generate resentment throughout the rest of the organization they were housed in: What gave those folks over there the seemingly unearned privilege to play in a sandbox? In fact, the playground analogy is incomplete. A more complete analogy is of research as a competitive sport. And by competitive sport we do not mean one with a "winner takes all" mentality. We mean one characterized by the desire to perform at the highest levels and to be inspired by others' high performance. The slogan of the modern Olympic Games, attributed to Pierre de Coubertin, sets the bar:

> The most important thing in the Olympic Games is not to win but to take part, just as the most important thing in life is not the triumph but the struggle. The essential thing is not to have conquered but to have fought well.

Research as a competitive sport can and should be exhilarating and fun, can and should involve friendship and collaboration, but must also involve competition. As VN articulated of his experience (Narayanamurti & Odumosu, 2016, pp. 83–84):

> Bell Labs was . . . an intensely competitive environment. The clearest example of how this was fostered was in the yearly review process. In the early days, every single member of the labs was ranked in order to identify the top 10 percent and

the bottom 10 percent. As one of our interviewees said about appearing on the bottom 10 percent "if you got on that list a couple of years, you were kind of in big trouble and usually ended up working somewhere else. But I mean, it was a continuous process so every year you came up with it, there were always some people at the bottom no matter what was going on." And another said "everyone knew that it was kind of a competitive thing, survival of the fittest sort of thing. And you knew you had to face that, that you couldn't kind of slack off." This competition was partly driven by the extreme level of excellence represented by the research team at Bell. As an interviewee said "I could walk down the hall and find an expert in anything. I could find collaborators who would help me with anything. And that was about as good as it gets." Institutionally, there were occasionally multiple groups working on the same or similar problems albeit using different approaches. This redundancy generated a different kind of competition—competition among teams. A great achievement of Bell Labs was the fine balance it struck between competition and collaboration. One of our interviewee's captured it well: "I think you could think of it as a big family, a lot of sibling rivalry, a very successful big family. . . . There's a certain bonding that occurs that is really rather special. . . . I think it comes from that notion of it like being in a big family. There's some sibling rivalry, but everybody is doing fantastically well.

Key to competition is a means of evaluating performance so that one knows how well one is doing relative to one's peers and even relative to one's own prior performance—and thus to hold research accountable. It is the opposite of anonymity, where poor performance can hide behind process or lack of knowledge by research leadership. Research accountability unleashes several benefits.

First, research performance continually improves because it is continually evaluated, with poor performance continually weeded out. Along these lines, it is critical that research be open and published so that performance is compared to that at the highest levels worldwide. If research is only communicated internally, it becomes easy to "drink one's own Kool-Aid" without being aware of excellent work going on outside and without being aware of the mistakes that the world will find that you couldn't find. For research this is especially important because, in exploring the unknown, mistakes *will* be made. Sunlight is the best disinfectant. Research must face outward first, inward second, if it is to be competitive.

A second benefit of research accountability is that process experimentation is unleashed. The process of nurturing research need not be static, rather processes can be dynamic and the subject of experimentation. Though this chapter articulates guiding principles for nurturing research, surely these principles should continue to evolve as we learn. For example, understanding better the correlation between research results and the characteristics of researchers who contribute such results should enable continuous improvement in the processes of recruiting, hiring, and retaining.

Third, collaborative, interdisciplinary work and appreciation for diverse perspectives naturally emerge without having to be forced. Competition incentivizes collaboration because collaborative and interdisciplinary work is a key route to research success. Those who do not collaborate and do not seek diverse perspectives are disadvantaged in the pursuit of research excellence.

And finally, research accountability introduces an element of urgency. Without accountability, research can easily become complacent. Instead, researchers must earn the privilege to do research by being held accountable for accomplishing outstanding research. Research is not a free lunch; it is a privilege, not a right. Note that this is particularly important for research—complacency is rarely a problem for development because it always has clear accountability to the metric of utility, which is more clear-cut than the metric of surprise.

To what do we hold researchers and research leadership accountable? The simple question asked at the end of a performance review cycle might be "What contributions have you made to actual or potential-future advances in useful learning and surprise over the past review cycle?" The contributions will, of course, be different at the different levels of the organization, but all levels must have "skin in the game," and all levels must be held accountable.

For researchers, accountability means contributing to surprise and use-fulness via any or all of the mechanisms of the technoscientific method. And, by "usefulness," we mean the broad definition discussed in Chapter 1—whatever humans desire, whether the desire is practical and satis-fies more immediate human needs or whether the desire is to satisfy human curiosity. Note that we don't hold researchers accountable for the various characteristics that one might use to select researchers to recruit and hire. It is for research *results* that researchers must be primarily held accountable and only secondarily for qualities that might or might not be leading indicators of potential for research results. Evaluation must primarily be done in hindsight, after the work has been done, not for proposals for work to be done. That said, secondary factors must carry some weight, because research results depend too much on historical contingency and luck. As articulated by Ralph Bown, vice president of research at Bell Labs from 1951 to 1955 (Bown, 1953, p. 4):

> A conviction on the part of employees that meritorious perfor-mance will be honestly appraised and adequately rewarded is a necessary ingredient of their loyalty. This appraisal, to be fair and convincing, must be based on the individual's performance and capabilities rather than wholly on the direct value of his re-sults. A system which rewards only those lucky enough to strike an idea which pays off handsomely will not have the cooperative teamwork needed for vitality of the enterprise as a whole.

Evaluation must consider paths that are not immediately successful, that show initial poor performance but have potential for long-term signifi-cant impact. Evaluation depends not just on final performance but also on the path of performance. Some of those paths may involve repeated failure but often involve learning. Indeed, at Bell Laboratories, the free-dom to fail was embedded in its culture along with the patience to suc-ceed. As VN describes (Narayanamurti & Odumosu, 2016, pp. 81–82):

> Perhaps the greatest distinction between the research orienta-tion at Bell Labs and research cultures elsewhere was the free-dom to fail. Members of the research staff enjoyed this freedom

in its totality. . . . This also meant a lot of patience on the part of the institution with projects that were initially failures but promised extraordinary returns if proven successful. As one of our interviewees said, "they were given the freedom and the money and encouragement to work on that for six years. You won't find that today. Al Cho, who developed molecular beam epitaxy, I think he worked over ten years before he grew good crystals. Again, that was expensive." The freedom to fail sometimes meant the patience necessary to succeed.

Most importantly, evaluation must be *fair*. It must not favor the established researcher over the younger researcher—always the young must be nurtured. Those doing the evaluations have a great burden. Implicit bias must be guarded against, data and diversity of opinions must be embraced, and human judgment must always be scrutinized even as it is, in the end, what we must rely on.

Research Leadership

For research leadership, accountability means contributing to the implementation of all three of the overarching guiding principles: aligning organization, funding, and governance for research; embracing a culture of holistic technoscientific exploration; and nurturing people with care and accountability. But it should not be underestimated how difficult these guiding principles are to implement in practice. Research is a very human endeavor: it does not happen just by abstract guiding principles; it happens by the human *and* intellectual contact between particular research leaders, particular researchers, and particular research directions.

For example, as discussed earlier, research leadership has the important responsibility of *jointly* nurturing researchers and research directions as both evolve in new ways—the building of critical mass. It is far easier to allow researchers to advance in whatever direction they are interested in, to respond to whatever proposal call has come in. To do the necessary joint nurturing requires research leaders with technical vision, judgment, competence, and empathy. It does not mean that research leaders know as much as researchers about researchers' particular specialties—that is, of course, impossible. But it does mean that research leaders should be technically proficient enough to understand their researchers' work and its importance. Research leaders who grew up in the research ranks, who themselves contributed significantly in their early careers to research, who have a visceral and intuitive feel for research, and who are respected technically by researchers—these are not just "nice to haves"; these are "must haves." Based on his experience at Bell Labs, (Narayanamurti & Odumosu, 2016, pp. 87–88), VN states that the administration was composed of

> highly accomplished researchers who were flexible enough to allow for new kinds of collaborative organizations and could understand the technical and scientific merits of the work of their teams. Their entire orientation was to help, serve, and assist their group members to obtain the necessary resources to be as successful as they could be. Finally, the administration was also capable of discussing the technical merits of various projects and determining their value. These determinations were not always correct, and sometimes accomplished researchers didn't always make good administrators, but a large number of them did amazingly well and their success was in no small part responsible for the enabling research environment at Bell.

Performance Review

For both researchers and research leadership, the practical instantiation of accountability is performance review. The most powerful messages on what the organization values are disseminated through the performance review process. Generally, you want people to know that the organization is a place where taking risks in order to seek surprise can be rewarded, not merely tolerated; where seeking "what you don't know" rather than making use of "what you already know" can be rewarded; where publishing and sharing rather than hoarding knowledge can be rewarded. In research, one cannot always wait for all i's to be dotted and t's to be crossed. Performance review is where researchers are accountable for contribution to the metagoal of surprise and creative destruction, and where research leadership is accountable for contribution to the three guiding principles for nurturing research. Moreover, the relationship between researchers and research leadership is a two-way street, with the leader's own performance review just as important as the researcher's performance review. Indeed, a poor performance review for a researcher reflects somewhat on the leader, in how well they have nurtured the researcher.

Note that we mean contribution in the broadest possible sense—not just to one's own research. Research and research leadership are sometimes solo efforts, in which case one's contribution to research outcome is more easily evaluated. Research and research leadership are more often collaborative efforts, however, in which case any one person's contribution to research outcome is less easily identified and evaluated. Every effort, thus, must be made to evaluate contribution generously and fairly so as to encourage collaboration. It is worth remembering that such an evaluation process is very different from those in typical academic institutions whose junior faculty are too often incentivized to emphasize individual contributions over collaborative activity to strengthen their case for unique work output.

This is not to denigrate solo efforts: whether an effort is best done solo or in collaboration is situational and depends on the topic area, its state of maturity, the absence / presence of suitable collaborators, and so on. But, on average, collaborations, when appropriate, more often lead to impacts that are much greater than the sum of their parts than do solo efforts. Thus, accountability for contribution to high-impact results or outcomes will drive impactful collaborations, provided performance evaluations fairly and generously consider contributions to collaboration. Researchers and research leaders who work in such a collaborative culture view seeking help from colleagues as natural, regardless of whether providing such help is within their colleagues' formal job descriptions. In this kind of culture, researchers and research leaders do not need to be equally good at all facets of the job. Some researchers are more articulate and better at writing papers; others are more comfortable solving problems in the lab. Some research leaders are more skilled at selecting those with potential to become elite researchers, and it would be reasonable for these managers to be preferentially more responsible for the larger organization's recruiting and hiring. A signature of a great research institution is the degree to which extreme talents, expertise, and interests can not only be accommodated but can flourish.

This is also not to lurch to the other extreme of purely committee work. Such efforts easily devolve into collaborations with no accountability—where committee members "hide" behind the committee that makes the decision. Always there must be individual human judgment, with committees a good source of advice but not making the final decision.

Because research and research leadership are complex human endeavors, they must be judged by humans. Processes must be put in place to enable as much fairness and objectivity as possible, including the use of diverse quantitative and qualitative input. Conferences, invited talks, being sought after, and other forms of external recognition are important indicators of research excellence. Bibliometric data (such as h-indexes) and peer review assessments are also valuable. But bibliometric data can skew toward "fast following," and peer review can skew toward conventional wisdom, not informed contrariness. Research leaders should take quantitative metrics and peer reviews into account but must counterbalance these with their own intellectual assessments of contribution to surprise. Leaders should be expected to attend major technical conferences to be in touch with the research frontiers, to know the best people in the field, and to know how their own people compare. If a research leader is unable to make sound assessments, the solution is not to substitute quantitative metrics or to soften the evaluation; the solution is to find new research leaders. And if a research leader's administrative duties become too time-consuming, the solution is not to stop attending conferences or connecting with researchers; the solution is to reduce those duties or to bring on board additional administrative support, always with the leader providing deeply interactive oversight.

Importantly, performance reviews must include both positive and negative feedback.

On the positive feedback side, many options are available.

The first reward for excellence in research or research leadership is local research environment: increased freedom and resources to perform and lead research. The exact form that this takes is situational and depends on how the researcher or research leader best performs. It can be in the form of being relieved from the distractions of administrative duties that often accompany success in research or research leadership. It can be in the form of increased support for laboratory equipment. It can be in the form of increased support for postdocs and students. It can be in the form of a longer leash into research directions that are further afield from the organization's overarching research direction. For many, if not most, researchers and research leaders, an organization that creates an environment optimal for the individual's and group's research productivity is the single most important reward.

The second reward is monetary: increased salary. This second reward, though not as important as the first, is not unimportant. Researchers and research leaders have personal, not just professional, lives, and as their personal lives are enriched and made more satisfying, so their professional lives will also be made easier and more satisfying. Salary is not the only measure of monetary reward, but it plays a significant role. It is a judgment call as to what salary differentials balance the enrichment of personal and professional lives without corroding egalitarian *esprit de corps*. Neither extreme is desirable. If an exorbitant salary is required to keep a researcher or research leader from moving, then it could be best for the researcher or research leader to move. But salaries that are not competitive will fail to attract the best researchers or research leaders. Most importantly, salaries and salary differentials must be determined by contribution to surprise and creative destruction, not by proposals won, by the size of one's research "empire," or by near-term impact on the mission of the larger institution. Researchers and research leaders must always have their eyes on long-term (not short-term) global (not local) technoscientific impact.

The third reward is reputational. Reputational reward is largely external—the result of excellence in research or research leadership recognized by the larger community of researchers or research leaders outside the organization. Such external reputational reward is partly an outcome of the research environment provided by the organization to researchers and research leaders, but organizations can also be influential in the encouragement of external society fellowship or awards, and appreciate those who take the time to nominate colleagues. Care must be taken for this to be done fairly, however, not overmarketing one's own people beyond their real contributions. Reputational reward can also be internal—promotions into various internal job categories. Such internal reputational rewards must be considered carefully, however. Because they are so visible, they can easily become a source of discontent and counter the sense of egalitarianism that is so important to collaboration and *esprit de corps*. Indeed, an overproliferation of job levels can be a dangerous sign that the organization is trying to compensate for the lack of a research environment that affords the opportunity to do or lead fulfilling research. When people feel their work supplies them with personal and professional satisfaction and purpose, which many researchers and research leaders feel when they are actually able to do or lead research, they don't feel the need to be compensated as much in salary, job status, or other perks. But when their work does not supply them with a higher purpose, perhaps because their research or research leadership is thwarted, they do need those other kinds of compensation. Doing what you love is what is most important, and for researchers and research leaders, it is research that they love.

On the negative feedback side, leadership has fewer options, most of them difficult and painful. Indeed, so painful that management at all levels will sometimes go to extremes to avoid it. Nonetheless, as difficult and painful as negative feedback is, it must be given. As articulated by Narayanamurti and Odumosu for researchers, but as is also applicable to research leadership (Narayanamurti & Odumosu, 2016, p. 91):

> Promotion and review at Bell Labs were done yearly, with a focus on quality not quantity. Individual researchers filled out single page self-assessments and then these were passed along to department heads who then came together and ranked everyone. The responsibility to promote and explain the work of individual researchers lay with the department heads who had to be very knowledgeable about the work of the people they led. This review process resulted in identifying people at the top, middle, and bottom of the pack. Using a roughly sliding five-year window, people at the bottom of the pack would feel pressure and in a few years be asked to move on unless their work improved. This constant pruning and categorizing resulted in a ubiquitous excellence. As one of our interviewees put it: "The average Bell Lab scientist was more often than not much better than people outside. It sounds arrogant, but I can tell you that that was the case."

The ultimate negative feedback is the realization that person and organization aren't matched, and organizational movement is necessary. Research or research leadership is not for everyone at all stages of their careers. But always this realization and its ramifications must be made with care, respect, and grace. Well before the realization, every effort must be made to mentor, to take leading indicators seriously, to give feedback, and to improve research performance. And if organizational movement is ultimately necessary, every effort should be made to find other organizations within the larger institution, or other institutions, that provide better matches. If the researcher or research leader was once an impactful contributor, they have undoubtedly developed capabilities in demand elsewhere, and every effort should be made to encourage a match with that demand.

In sum, performance review based on meritocratic competition is critical and must be based on actual performance in research or research leadership. Elite research is no different from elite performance in any walk of life—it does not come easy or for free—and this means that the performance review must have teeth. Outstanding researchers and research leaders should be encouraged to stay and be given more and more latitude, almost like tenure. But those who are less successful at research or research leadership must be encouraged to move on, or else the organization suffers from lower standards and morale. And, of course, those who are less successful at research or research leadership might very well be extremely successful at other pursuits.

Finally, note that it is not a sign of weakness in an organization when researchers or research leaders move on of their own volition, often due to the vagaries of personal and professional motivation and career choice. It is a sign of strength when researchers and research leaders in an organization *have* opportunities to move on, especially to external positions on the world stage of research. Likewise, it is a sign of weakness when researchers and research leaders do not have such opportunities. What is important is that the organization be able to attract top young researchers to fill a pipeline of talent to replace those who are moving on.

4.4 Recapitulation

In Chapter 4, we have articulated a rethinking of the nurturing of research, organized around a small number of guiding principles. The principles were intended to be general enough to apply to a wide range of research organizations, yet specific enough to be actionable.

The first principle, *align organization, funding, and governance for research,* acknowledges that research is a highly specialized activity, one that seeks learning and surprise, and one whose outcome cannot be predicted in advance. For that reason, it must be treated very differently from its equally important partner in knowledge production, development. Research must occur against an organizational, funding, and governance backdrop whose purposes, structures, resources, and leadership are aligned with research.

The second principle, *embrace a culture of holistic technoscientific exploration,* acknowledges that the alignment of organization, funding, and governance with research does not guarantee that the outcome of research will be surprise and creative destruction. A successful research culture embraces all the different mechanisms of technoscientific exploration discussed in our three previous chapters: from Chapter 1, embrace the full technoscientific method; from Chapter 2, embrace finding questions *and* answers; and, from Chapter 3, embrace the new, the surprising, and the informed contrarian.

The third principle, *nurture people with care and accountability,* acknowledges that people are the beating heart of research—they must be cared for but also held accountable to high standards. There is the empathy and care associated with recruiting and hiring, with joint nurturing of researchers and research directions, and with rewarding for more successful research outcomes. There is also the selectivity and accountability associated with not hiring, with weeding out research directions that are no longer at the research frontier or have morphed too far afield from the overarching mission of the organization, and with accountability for the highest standards of excellence in research outcome.

Conclusion

It takes only a moment of reflection to be astonished by the technoscientific revolutions that have remade human society during just this past century and a half. A few examples from the physical sciences and engineering have been discussed in this book: special relativity, the transistor effect, the light bulb, the transistor, the laser, the blue LED, and the iPhone. Some examples from the life and information sciences, though not discussed in this book, are DNA, the polymerase chain reaction method, CRISPR-Cas9 gene-editing tools, and deep learning. Who knows what technoscientific revolutions might remake human society in the future, at any level of the "more is different" seamless web of knowledge discussed in Chapter 2: physical science and engineering, life science and medicine, and social science and human affairs. We are certainly mindful of the positive and negative consequences of techno-scientific advance, and thus of humanity's collective responsibility for managing those consequences. As articulated by science and technology studies scholar Sheila Jasanoff (Jasanoff, 2020, p. 9):

> If engineering has emerged as the powerhouse of progress in the
> 21st century, with [that] power comes responsibility.

But we are confident in the limitlessness of the transformative and ulti-mately beneficial technoscience waiting to be created.

Just as technoscience shapes society, society also shapes, and must shape, research—the front end of the research-and-development cycle that is the genesis of technoscientific revolutions. How does society shape research? It does so through the social enterprises within which research is done. The first formal research organization, Thomas Edi-son's Menlo Park Laboratory, formed in 1876, was a social enterprise. The great industrial research laboratories of the twentieth century—including Bell Labs, IBM, Xerox PARC, Dupont, and GE—were social enterprises. Today's research universities, research institutes, and na-tional and international laboratories are social enterprises.

As society and its values have evolved, these social enterprises have evolved and, in turn, research has evolved. In the latter half of the twenti-eth century, following World War II, modern society's support for re-search was accelerated by its belief in the power of technoscience as a public and collective good, particularly for national defense, economic prosperity, and human health. Subsequently, in the latter quarter of the twentieth century continuing into the first quarter of the twenty-first century, modern society began to shift its emphasis to short-term and narrower measures of return on capital invested and to a transactional "what's in it for me" approach to research. The consequence was that the great industrial research laboratories shifted their emphasis from re-search to development and, in some cases, eliminated research entirely. As we approach the second quarter of the twenty-first century, our hope is that modern society will move away from viewing research as a trans-actional creator of short-term private benefit and toward a view of re-search as a creator of long-term public and *collective* benefit, addressing today's societal grand challenges and enabling society-transforming ad-vances we can now only dimly imagine.

In the foyer of the main entrance to Bell Labs, beneath the bust of the great inventor Alexander Graham Bell, is engraved this famous quotation:

> Leave the beaten track behind occasionally and dive into the woods. Every time you do, you will be certain to find something you have never seen before.

Many at Bell Labs, including one of us (VN), made it a tradition to bring visitors to honor this bust and its engraving (Narayanamurti, 1987). How research organizations can best leave the beaten track to find the new and surprising, however, isn't so simple—even with the "will" to unleash the full power of research and its technoscientific revolutions, few have found the "way." In this book, we have tried to emphasize that the "way" requires research to be nurtured, not simply managed, and that there are timeless and overarching principles associated with such nurturing.

Our hope is that such principles enable a new generation of "Research Labs 2.0" that go beyond our current generation of "Research Labs 1.0." This new generation of "Research Labs 2.0" could come in very different shapes and sizes. They could have different organizational governance structures, funding models, technoscientific knowledge domain foci, and scales—some at the large scale of CERN, some at the medium scale of a research organization embedded in a larger corporation, and some at the small-scale of a philanthropically supported research institute. But they would all be aware of the timeless and overarching principles associated with nurturing research.

This new generation of labs would be aware that nurturing research must be aligned with what is being nurtured. It must be aligned with the nature of research and all its feedback loops and amplifications—not just the reductionism of physics, though that is important, but also the complexity sciences and the fusion of disciplines with a holistic "more is different" attitude (American Academy of Arts & Sciences, 2013; National Research Council, 2014). If the symbiosis between science and technology is not understood, then the full technoscientific cycle and its collective power will not be embraced. If the symbiosis between question-finding and answer-finding is not understood, question-finding especially is fragile and will go unsupported. If the importance of surprise and the overturning of conventional wisdom is not understood, then informed contrarians who are always questioning, always pushing past conventional wisdom's comfort zone, will be weeded out.

This new generation of labs would also be aware that nurturing research requires an appreciation of research as a deeply human, collaborative, and social endeavor. Like all social endeavors, it requires active social construction: organization, funding, and governance that is aligned with research; a human culture that supports holistic technoscientific exploration; and the nurturing of people with care and accountability. Like all collaborative endeavors, it benefits from the full diversity and inclusiveness of the society in which it is embedded and to which it will contribute. Like all human endeavors, particularly those that require extraordinary performance, it requires nurturing the whole human being and spirit. We have borrowed his insights many times throughout this book but cannot help but borrow once more. To paraphrase the words of Ralph Bown, vice president of research at Bell Labs from 1951 to 1955 (Narayanamurti & Odumosu, 2016, p. 76),

> [R]esearch environments reflect human relationships and group spirit. In short, successful research institutions should never forget that they are human institutions and they should place people above structure.

Our hope is also that, in the more distant future, an increasingly effective generation of research laboratories, "Research Labs 3.0," will emerge. This new generation would be based on improved principles that go beyond, and perhaps even overturn or displace, the ones articulated in this book. To reach beyond where we find ourselves today, we must continue to "learn how to learn" (Odumosu et al., 2015). We recognize two sources of learning. First, real experiments: as we become more deliberate in designing research organizations and watching them operate, we must use these same organizations as experiments from which to learn more about the nature and nurturing of research. The nurturing of research is a body-contact sport that requires direct human experience. Second, artificial experiments: as artificial intelligence advances and begins to learn how to learn, we look forward to mapping the nature of artificial learning onto the nature of human learning. Much has changed in the social enterprise of research in the seventy-five years since Vannevar Bush's seminal report to President Roosevelt and the creation of the National Science Foundation. We believe at least as much change is possible in the coming seventy-five years. As articulated by President Joseph R. Biden, Jr. (Biden, 2021):

> I believe it is essential that we refresh and reinvigorate our national science and technology strategy to set us on a strong course for the next 75 years, so that our children and grandchildren may inhabit a healthier, safer, more just, peaceful, and prosperous world. This effort will require us to bring together our brightest minds across academia, medicine, industry, and government—breaking down the barriers that too often limit our vision and our progress, and prioritizing the needs, interests, fears, and aspirations of the American people.

To articulate this book's rethinking of the nature and nurture of research, we have, in some cases, used existing words and phrases in new or nuanced ways. We have collected them here along with our condensed and sometimes nuanced definitions, though we also encourage the reader to use the index to find the words and phrases in the body of the book for fuller context.

A / B Testing   A user-experience research methodology in which users (especially web-page users) are presented with two variants and their responses used to deduce what they prefer—what "questions" they would like "answered."

Adjacent Possible   The space of latent questions and answers enabled by combining questions and answers that already exist in "the possible."

Ansatz   In physics and mathematics, an educated or inspired guess of some underlying assumption that helps answer a question (explain some fact) or solve a problem (find a form that fulfills some function).

Answer-Finding   The finding of new answers or solutions to existing questions or problems.

"Applied" Research   A phrase that has come to mean research of a technological flavor. In our view, this phrase, along with the phrase "basic research," is limiting and should be retired. The adjective "applied" carries with it the sense of "less foundational," and thus the phrase "applied research" conflates research of a technological flavor with less foundational research. In fact, research of a technological flavor (engineering research) is just as foundational. The degree to which knowledge advance is more foundational does not depend on its scientific or technological flavor.

It depends solely on its degree of surprise—the degree to which it contradicts and forces a rethinking of conventional wisdom.

Auxiliary Problem   A problem that, if it could be solved, would enable solving another problem of more immediate interest.

"Basic" Research   A phrase that has come to mean research of a scientific flavor. In our view, this phrase, along with the phrase "applied research," is limiting and should be retired. The adjective "basic" carries with it the sense of "more foundational," and thus the phrase "basic research" conflates scientific research with more foundational research. In fact, research of a technological flavor (engineering research) is just as foundational. The degree to which knowledge advance is more foundational does not depend on its scientific or technological flavor. It depends solely on its degree of surprise—the degree to which it contradicts and forces a rethinking of conventional wisdom.

Bricolage   In our context, the combining of knowledge modules "as is" without internal modification of the modules to improve their fit into the combined whole. See also **Multidisciplinary**.

Complex Adaptive Systems   Systems whose internal structures modify to enable success in their environment, which can be either biological or engineered, whose hierarchical and modular order facilitate adaptation, and whose goal-directed adaptations create nonrandom substructures. Human technoscientific knowledge is a quintessential complex adaptive system.

Confirm Belief   Conventional wisdom "believes" that potential new knowledge will be useful and, after further playing out, the knowledge is indeed found to be useful.

Confirm Disbelief   Conventional wisdom "disbelieves" that potential new knowledge will be useful and, after further playing out, the knowledge is indeed found not to be useful.

Consolidation   See **Learning by Surprise and Consolidation**.

Conventional Wisdom   The body of accepted sociocultural technoscientific paradigms and the technoscientific knowledge on which those paradigms draw.

Creative Destruction   In the context of this book, surprising advances in knowledge create new paradigms, but, because they are surprising, also destroy previous paradigms and overturn conventional wisdom.

Culture   In this book, the human values, desires, interests, norms, and behaviors that influences what $\dot{S}$ and $\dot{T}$ society or a research organization invests in, and how that $\dot{S}$ and $\dot{T}$ is "done."

Curiosity   See **Utility.**

Development   The back end of the research and development (R&D) cycle that advances technoscientific knowledge. Successful development outcome is characterized by consolidation—the consolidation of conventional wisdom and the extension of existing sociocultural technoscientific paradigms. Development can have scientific and / or technological flavors.

Disconfirm Belief   Conventional wisdom "believes" that potential new knowledge will be useful but, after further playing out, the knowledge is found not to be useful.

Disconfirm Disbelief   Conventional wisdom "disbelieves" that potential new knowledge will be useful but, after further playing out, the knowledge is indeed found to be useful.

Discovery   The quintessential outcome of research of a scientific nature—a surprising new fact or a surprising new explanation for a fact.

"Discovery" Research   In our nomenclature, this is simply research—it is the exploratory front end of the research-and-development cycle that is driven by the metagoal of surprise. See also **"Foundational" Research.**

Engineering Method   The three mechanisms that occur cyclically to advance technological knowledge: function-finding, form-finding, and exapting. The combination of the engineering method with the scientific method is what we call the technoscientific method.

Engineering Research   Research with a technological flavor.

Exapting   The finding of new function from existing form. The repurposing or co-opting, often opportunistic and serendipitous, of existing forms for new function. This is one of the six mechanisms of the technoscientific method.

Explanation-Finding   The finding of explanations for facts about the world or of deeper explanations for shallower explanations of facts about the world. This is one of the six mechanisms of the technoscientific method.

Fact-Finding   The finding of facts: stable observed patterns about the world that are of human interest and for which explanations might be found. This is one of the six mechanisms of the technoscientific method.

Form-Finding   The finding of forms (artifacts and the processes by which those artifacts are made and used) that fulfill human-desired functions. This is one of the six mechanisms of the technoscientific method.

"Foundational" Research   In our nomenclature, this is simply research—it is the exploratory front end of the research-and-development cycle that is driven by the metagoal of surprise. We call it "foundational" because it leads to the creation of new paradigms that become foundational to the subsequent extension of those paradigms and the subsequent consolidation and strengthening of the associated emerging conventional wisdom.

Function-Finding The finding of human-desired functions. This is one of the six mechanisms of the technoscientific method.

Generalizing The repurposing of explanation, created to explain one set of facts or shallower explanation, to explain other facts or other shallower explanation, including to "hypothesis test" the explanation. This is one of the six mechanisms of the technoscientific method.

Gestalt An organized whole that is perceived as more than the sum of its parts.

Hail Mary A desperate attempt with only a small chance of success, based on the so-called "Hail Mary" pass in American football. In the context of this book, it is an *uninformed* contrarian idea whose potential utility runs contrary to conventional wisdom but is not informed by deep thinking or intuition, so it relies mostly on luck for its success.

Hourglass A fixed protocol that facilitates a rich evolution of knowledge modules on both sides of the protocol. An exemplar is the Edison socket. As long as the mechanical and electrical specifications of the Edison socket were respected, both a rich evolution of electricity-generating technologies (natural gas generation, nuclear power, photovoltaics) for feeding electricity into the socket and a rich evolution of light-generating technologies (incandescent, halogen, solid-state lighting) could be accommodated.

Hypothesis-Finding A specific type of generalizing. Generalizing takes existing theory and finds facts that are predicted by, but were not originally used to construct, the theory. Some of those facts might have been observed already, in which case the theory is immediately made more plausible. Some of those facts might not have been observed previously—these predicted-but-not-yet-observed facts represent the finding of a hypothesis to be tested by experimental observation.

Hypothesis-Testing The testing of a hypothesis about the world: the attempt to observe predicted-but-not-yet-observed facts about the world. If the fact is not observed, then the theory on which the hypothesis is based is falsified or at least made less plausible. If the fact is observed, then the theory on which the hypothesis is based is made more plausible.

Implausible Utility A proposed knowledge advance whose utility is implausible to conventional wisdom but whose utility is ultimately verified.

Informed Contrarian   A researcher or research group whose proposed knowledge advance is contrary to conventional wisdom but ultimately verified.

Interdisciplinary   The combining of knowledge modules from different knowledge domains that do not work "as is" but rather require internal adjustment and refinement of the knowledge modules to enable them to fit together. Successful combination thus requires both breadth *and* depth—breadth to create the combinations and depth to make the combinations work.

Invention   The quintessential outcome of engineering research: a surprising new function and / or a surprising new form that fulfills an existing function. The first step in the innovation chain that ends in broad societal use.

Knowledge Modules and Submodules   A knowledge module is a set of closely related question-and-answer pairs that "fit" together. It might in turn be composed, in a hierarchical manner, of knowledge submodules or subsets of question-and-answer pairs that themselves "fit" together and integrate up to the larger knowledge module. A scientific knowledge module might be an explanation composed of ever deeper explanations. A technological knowledge module might be a component built from subcomponents that themselves are built from subsubcomponents.

Knowledge Spillover   Knowledge produced at some cost by one party that benefits ("spills over to") another party who does not pay that cost. Knowledge is prone to this because it is a "nonrival" good—the consumption of the good by one party does not reduce the amount available for other parties.

Learning by Surprise and Consolidation   We distinguish between two kinds of learning. One kind, learning by learning by surprise, involves the overturning of conventional wisdom; the other kind, learning by consolidation, involves the consolidation and strengthening of conventional wisdom.

"Linear" Model   The model of the evolution of technoscientific knowledge in which science leads to technology and then to application. This simplistic model is manifestly incorrect, in that science and technology feed on each other in cycles and in no particular order. See also **"Basic" Research** and **"Applied" Research**.

Multidisciplinary   The combining and use of knowledge modules from different knowledge domains "as is," without refinement, to enable them to fit better together. The combining and use require breadth, but not necessarily depth, of knowledge. See also **Bricolage**.

Next-Adjacent Possible   The space of latent questions and answers enabled by combining questions and answers that are themselves latent in the "adjacent possible," and that, in turn, are enabled by combining questions and answers that already exist in the "possible."

"Normal" Science   The consolidation of conventional wisdom, and the extension and strengthening of paradigms, with a science flavor. This is scientific development, analogous to engineering development (standard engineering) but with a science flavor.

Ockham's Razor   The idea, attributed to English Franciscan friar William of Ockham (1287–1347), that the simpler and more parsimonious an explanation of observational patterns, the more likely the explanation to be true. Parsimonious explanations are more generalizable to predict other patterns, hence more easily falsifiable if the prediction does not hold and, if not falsified, are more likely to be true.

Paradigm   Holistic combinations of knowledge "put to work" to advance knowledge: a kind of metaknowledge of how to use knowledge modules to accomplish dynamic changes to knowledge modules. Paradigms are combined sociocultural and technoscientific constructions, binding a social community of technoscientists together in common agreement on how to move technoscientific knowledge forward.

Paradigm Creation   The quintessential outcome of research: the creation of new paradigms that surprise and overturn conventional wisdom and previous paradigms.

Paradigm Extension   The quintessential outcome of development: the extension of existing paradigms that consolidate and extend conventional wisdom.

Possible   The space of existing technoscientific questions and answers.

Practical Application   See **Utility**.

Problem-Finding   We take problem-finding for technological knowledge to be the equivalent of question-finding in scientific knowledge, and, for shorthand, we take question-finding to encompass both question-finding and problem-finding.

Punctuated Equilibria   Using language borrowed from evolutionary biology: the process of knowledge evolution in which periods of continuous advance and "consolidation" of conventional wisdom are punctuated by discontinuous advance and "surprise" to conventional wisdom. See also **Conventional Wisdom** and **Learning by Surprise and Consolidation**.

"Pure" Research   A phrase taken to mean research inspired by human curiosity. In our view, this phrase should be retired. The adjective "pure" carries with it, not only the sense of being uninspired by considerations of practical application, but also the sense of not being able to impact practical application. Though this conclusion may be true for development, it is by no means true for research. Research is characterized by the meta-goal of surprise. Surprise means that the particular knowledge advance cannot be anticipated and, all the more, that the subsequent downstream impact of that knowledge advance cannot be anticipated. Inspiration by any means is not a good leading indicator of the ultimate impact of research.

Question-Finding   The finding of new questions or problems of human interest to answer or solve.

Research   The front end of the research-and-development (R&D) cycle that advances technoscientific knowledge. Successful research outcome is characterized by surprise: the overturning of conventional wisdom and the creation of new sociocultural technoscientific paradigms. Research can have scientific and / or technological flavors.

Science   The static repository (S) of facts and explanations for those facts. Sometimes conflated with the "doing" of science ($\dot{S}$), which is the scientific R&D that advances science.

Scientific Method   The three mechanisms that occur cyclically to advance scientific knowledge: fact-finding, explanation-finding, and generalizing. The combination of the scientific method with the engineering method is what, in this book, we call "the technoscientific method."

Solution-Finding   We take solution-finding for technological knowledge to be the equivalent of answer-finding in scientific knowledge, and for shorthand we take answer-finding to encompass both answer-finding and solution-finding.

"Standard" Engineering   The consolidation of conventional wisdom and the extension and strengthening of paradigms, with a technological flavor. This is engineering development, analogous to scientific development ("normal" science) but with a technological flavor.

Stylized Fact   A term borrowed from the social sciences, referring to empirical findings that are true, general, and important, and which can thus be accepted as true and worthy of explanation-finding.

Surprise   See **Learning by Surprise and Consolidation**.

T-Shaped Research / Researcher   Researchers or research organizations who combine depth and breadth of knowledge and can thus engage in interdisciplinary research.

Technology The static repository (T) of human-desired functions and forms that fulfill those functions. Sometimes conflated with the "doing" of technology ($\dot{T}$), which is the engineering R&D that advances technology.

Technoscience The sum of science and technology, of scientific and technological knowledge.

Technoscientific Method The combination of the scientific and engineering methods and their interactions.

Useful Learning The degree to which an advance in knowledge provides both utility and learning. Development values utility more highly, while research values learning more highly. The greater the useful learning, the greater the advance in knowledge, with the epitome of creativity being when both utility and learning are extreme.

Utility (Curiosity and Practical Application) An advance in scientific or technological knowledge has utility if it has an impact on, or satisfies, human interests or desires. In human interests or desires, we include both human curiosity (such as discovering the meaning of human consciousness or inventing a faster bicycle just to see how fast it can ride) as well as practical applications (such as to eat).

REFERENCES

Adner, R., & Levinthal, D. A. (2002). The emergence of emerging technologies. *California Management Review, 45*(1), 50–66.

American Academy of Arts & Sciences. (2013). ARISE II: Advancing research in science and engineering: Unleashing America's research and innovation enterprise.

American Academy of Arts and Sciences. (2014). Restoring the foundation: The vital role of research in preserving the American dream. https://www.amacad.org/publication/restoring-foundation-vital-role-research-preserving-american-dream.

Anderson, P. W. (1972). More is different. *Science, 177*(4047), 393–396.

Anderson, P. W. (2001). Science: A "dappled world" or a "seamless web"? *Studies in History and Philosophy of Modern Physics, 32*(3), 487–494.

Anderson, P. W. (2011). *More and different: Notes from a thoughtful curmudgeon:* World Scientific.

Andriani, P., & Cattani, G. (2016). Exaptation as source of creativity, innovation, and diversity: Introduction to the special section. *Industrial and Corporate Change, 25*(1), 115–131.

Arrow, K. (1962). Economic welfare and the allocation of resources for invention. In Committee on Economic Growth of the Social Science Research Council Universities–National Bureau Committee for Economic Research (Ed.), *The rate and direction of inventive activity: Economic and social factors* (pp. 609–626). Princeton University Press.

Arthur, W. B. (2009). *The nature of technology: What it is and how it evolves:* Simon and Schuster.

Avina, G. E., Schunn, C. D., Odumoso, T., Silva, A. R., Bauer, T. L., Crabtree, G. W., Johnson, C. M., Picraux, S. T., Sawyer, R. K., Schneider, R. P., Sun, R.,

Feist, G. J., Narayanamurti, V., & Tsao, J. Y. (2018). The Art of research: A divergent / convergent thinking framework and opportunities for science-based approaches. In Eswaran Subramanian, J. Y. Tsao, & Toluwalogo Odumoso (Eds.), *Engineering a Better Future.* Springer, Chapter 14.

Azoulay, P., Graff Zivin, J. S., & Manso, G. (2011). Incentives and creativity: Evidence from the academic life sciences. *The RAND Journal of Economics, 42*(3), 527–554.

Bardeen, J. (1957). Research leading to point-contact transistor. *Science, 126*(3264), 105–112.

Bartolomei, H., Kumar, M., Bisognin, R., Marguerite, A., Berroir, J.-M., Bocquillon, E., Plaçais, B., Cavanna, A., Dong, Q., & Gennser, U. (2020). Fractional statistics in anyon collisions. *Science, 368*(6487), 173–177.

Basalla, G. (1988). *The evolution of technology:* Cambridge University Press.

Basov, N., & Prokhorov, A. (1955). About possible methods for obtaining active molecules for a molecular oscillator. *Начало лазерной эры в СССР,* 28.

Biden, J. R., Jr. (2021, January 15). Letter to Eric S. Lander, incoming director of the Office of Science and Technology Policy. https://www.whitehouse.gov /briefing-room/statements-releases/2021/01/20/a-letter-to-dr-eric-s-lander -the-presidents-science-advisor-and-nominee-as-director-of-the-office-of -science-and-technology-policy/.

Birks, J. B., & Rutherford, E. (1962). Rutherford at Manchester. In John Bettely Birks (Ed.), *An account of the Rutherford Jubilee International Conference at Manchester, Sept. 1961. With articles by Rutherford and others.* Heywood and Company.

Bloembergen, N. (1956). Proposal for a new type solid state maser. *Physical Review, 104*(2), 324.

Bowers, B., & Anastas, P. (1998). *Lengthening the day: A history of lighting technology.* Oxford University Press.

Bown, R. (1953). *Vitality of a research institution and how to maintain it.* Paper presented at the 1952 Conference on Administration of Research. Georgia Institute of Technology.

Braun, A., Braun, E., & MacDonald, S. (1982). *Revolution in miniature: The history and impact of semiconductor electronics.* Cambridge University Press.

Brinkman, W. F., Haggan, D. E., & Troutman, W. W. (1997). A history of the invention of the transistor and where it will lead us. *IEEE Journal of Solid-State Circuits, 32*(12), 1858–1865.

Brooks, Jr., F. P. (1996). The computer scientist as toolsmith II. *Communications of the ACM, 39*(3), 61–68.

Buchanan, M. (2003). *Nexus: Small worlds and the groundbreaking theory of networks:* WW Norton and Company.

Bush, V. (1945). Science, the endless frontier. A report to the president by Vannevar Bush, director of the Office of Scientific Research and Development, July 1945.

Campbell, D. T. (1960). Blind variation and selective retentions in creative thought as in other knowledge processes. *Psychological Review, 67*(6), 380.

Carter, J. (1981). State of the union address. https://www.jimmycarterlibrary.gov/assets/documents/speeches/su81jec.phtml#:~:text=January%2016%2C%201981,peace%20for%20four%20uninterrupted%20years.

Cech, T. R., & Rubin, G. M. (2004). Nurturing interdisciplinary research. *Nature Structural and Molecular Biology, 11*(12), 1166–1169.

Cheung, N., & Howell, J. (2014). Tribute to George Heilmeier, inventor of liquid crystal display, former DARPA director, and industry technology leader. *IEEE Communications Magazine, 52*(6), 12–13.

Cho, A. Y. (1971). Film deposition by molecular-beam techniques. *Journal of Vacuum Science and Technology, 8*(5), S31–S38.

Christensen, C., & Raynor, M. (2013). *The innovator's solution: Creating and sustaining successful growth.* Harvard Business Review Press.

Close, F. (2014). *Too hot to handle: The race for cold fusion.* Princeton University Press.

Darwin, C. (1859). *On the origin of species by means of natural selection.* John Murray.

Dasgupta, P., & David, P. A. (1994). Toward a new economics of science. *Research Policy, 23*(5), 487–521.

Dew, N., Sarasvathy, S. D., & Venkataraman, S. (2004). The economic implications of exaptation. *Journal of Evolutionary Economics, 14*(1), 69–84.

Dingle, R., Gossard, A. C., & Stormer, H. L. (1979). USA Patent No. 4,163,237. High mobility multilayered heterojunction devices employing modulated doping. https://patft.uspto.gov/netacgi/nph-Parser?Sect1=PTO1&Sect2=HITOFF&d=PALL&p=1&u=%2Fnetahtml%2FPTO%2Fsrchnum.htm&r=1&f=G&l=50&s1=4,163,237.PN.&OS=PN/4,163,237&RS=PN/4,163,237.

Dingle, R., Störmer, H., Gossard, A., & Wiegmann, W. (1978). Electron mobilities in modulation-doped semiconductor heterojunction superlattices. *Applied Physics Letters, 33*(7), 665–667.

Einstein, A. (1917). Zur quantentheorie der strahlung. *Zeitschrift für Physik, 18,* 121–128.

Einstein, A. (1934). On the method of theoretical physics. *Philosophy of Science, 1*(2), 163–169.

Einstein, A., & Infeld, L. (1971). *The evolution of physics.* Cambridge University Press Archive (first published 1938).

Farmer, J. D., & Lafond, F. (2016). How predictable is technological progress? *Research Policy, 45*(3), 647–665.

Feynman, R. (1974). Cargo cult science: CalTech 1974 Commencement Address. https://calteches.library.caltech.edu/51/2/CargoCult.htm.

Feynman, R. P. (2015). *The quotable Feynman.* Princeton University Press.

Feynman, R. P., Leighton, R. B., & Sands, M. (2011). *Six not-so-easy pieces: Einstein's relativity, symmetry, and space-time.* Basic Books (originally published 1963).

Franklin, B. (1791). *The autobiography of Benjamin Franklin*. Yale University Press (2003 Edition).

Funk, J. L. (2013). *Technology change and the rise of new industries*. Stanford University Press.

Galison, P. (1997). *Image and logic: A material culture of microphysics*. University of Chicago Press.

Garud, R. (1997). On the distinction between know-how, know-what, and know-why. *Advances in Strategic Management, 14,* 81–102.

Gertner, J. (2013). *The idea factory: Bell Labs and the great age of American innovation*. Penguin.

Godin, B. (2006). The linear model of innovation: The historical construction of an analytical framework. *Science, Technology and Human Values, 31*(6), 639–667.

Goudsmit, S. (1972). Editorial: Criticism, acceptance criteria, and refereeing. *Physical Review Letters, 28*(6), 331–332.

Gould, S. J. (1991). Exaptation: A crucial tool for an evolutionary psychology. *Journal of Social Issues, 47*(3), 43–65.

Gould, S. J., & Eldredge, N. (1993). Punctuated equilibrium comes of age. *Nature, 366*(6452), 223–227.

Gould, S. J., & Vrba, E. S. (1982). Exaptation—a missing term in the science of form. *Paleobiology, 8*(01), 4–15.

Graham, L. (2013, November 13). How we barely beat Soviet Russia to inventing the laser. *Gizmodo*.

Haitz, R., & Tsao, J. Y. (2011). Solid-state lighting: 'The case' 10 years after and future prospects. *Physica Status Solidi (a), 208*(1), 17–29.

Hammack, W. S., & DeCoste, D. J. (2016). *Michael Farady's the chemical history of a candle*. Articulate Noise Books (after original 1865 publication).

Harvey, J. E., & Forgham, J. L. (1984). The spot of Arago: New relevance for an old phenomenon. *American Journal of Physics, 52*(3), 243–247.

Hecht, J. (2010). A short history of laser development. *Applied Optics, 49*(25), F99–F122.

Hills, T. T., Todd, P. M., Lazer, D., Redish, A. D., Couzin, I. D., & Group, C. S. R. (2015). Exploration versus exploitation in space, mind, and society. *Trends in Cognitive Sciences, 19*(1), 46–54.

Hof, R. (2014, February 27). Peter Thiel's advice to entrepreneurs: Tell me something that's true but nobody agrees with. *Forbes*.

Hollingsworth, J. R. (2003). Major discoveries and excellence in research organizations. *Max-Planck Forum*, 215–228.

Hooker, S. (2020). The hardware lottery. *arXiv preprint arXiv:2009.06489*.

Hughes, T. P. (1987). The evolution of large technological systems. In Wiebe E. Bijker, Thomas P. Hughes, & Trevor J. Pinch (Eds.), *The social construction of technological systems: New directions in the sociology and history of technology*. MIT Press.

Hughes, T. P. (1993). *Networks of power: Electrification in Western society, 1880–1930*. Johns Hopkins University Press.

Jacob, F. (1977). Evolution and tinkering. *Science, 196*(4295), 1161–1166.

Jasanoff, S. (2020). Temptations of technocracy in the century of engineering. *The Bridge (National Academy of Engineering)*, 8–11.

Johnson, S. (2011). *Where good ideas come from: The natural history of innovation*. Penguin UK.

Kauffman, S. A. (1996). *At home in the universe: The search for the laws of self-organization and complexity*. Oxford University Press.

Kauffman, S. A. (2019). *A world beyond physics: The emergence and evolution of life*. Oxford University Press.

Kingsley, G. (2004). On becoming just another contractor: Contract competition and the management of science at Sandia National Laboratories. *Public Performance and Management Review, 28*(2), 186–213.

Kocienda, K. (2018). *Creative selection: Inside Apple's design process during the golden age of Steve Jobs*. St. Martin's Press.

Köster, M., Langeloh, M., & Hoehl, S. (2019). Visually entrained theta oscillations increase for unexpected events in the infant brain. *Psychological Science, 30*(11), 1656–1663.

Kroemer, H. (2001). Nobel Lecture: Quasielectric fields and band offsets: Teaching electrons new tricks. *Reviews of Modern Physics, 73*(3), 783.

Kuhn, T. S. (1974). Second thoughts on paradigms. *The Structure of Scientific Theories, 2,* 459–482.

Kuhn, T. S. (2012). *The structure of scientific revolutions* (4th ed.). University of Chicago Press (original edition published in 1962).

Laughlin, R. B., & Pines, D. (2000). The theory of everything. *Proceedings of the National Academy of Sciences, 97*(1), 28–31.

Layton, E. T., Jr. (1974). Technology as knowledge. *Technology and Culture,* 31–41.

Lewin, K. (1952). *Field theory in social science: Selected theoretical papers by Kurt Lewin*. Tavistock.

Lin, A. C. (2002). *Reform in the making: The implementation of social policy in prison*. Princeton University Press.

Lin, O. C. C. (2018). *Innovation and entrepreneurship: Choice and challenge*. World Scientific.

Mankins, J. C. (1995). Technology readiness levels. White paper, April 6, 1995. https://aiaa.kavi.com/apps/group_public/download.php/2212/TRLs_MankinsPaper_1995.pdf.

Marcus, G. (2018). Deep learning: A critical appraisal. *arXiv preprint arXiv: 1801.00631*.

Marx, M. R. S. L. (1994). *Does technology drive history?: The dilemma of technological determinism*. MIT Press.

Mazzucato, M. (2011). The entrepreneurial state. *Soundings, 49*(49), 131–142.

Merchant, B. (2017). *The one device: The secret history of the iPhone.* Random House.

Merton, R. K. (1973). *The sociology of science: Theoretical and empirical investigations.* University of Chicago Press.

Mimura, T., Hiyamizu, S., Fujii, T., & Nanbu, K. (1980). A new field-effect transistor with selectively doped GaAs / n-AlxGa1-xAs heterojunctions. *Japanese Journal of Applied Physics, 19*(5), L225.

Narayanamurti, V. (1987). On nurturing the novel neocortex: New VP shares views. *Sandia Lab News, 39.*

Narayanamurti, V., Anadon, L. D., & Sagar, A. D. (2009). Transforming energy innovation. *Issues in Science and Technology, National Academies, 26*(1), 57–64.

Narayanamurti, V., & Odumosu, T. (2016). *Cycles of invention and discovery.* Harvard University Press.

Narayanamurti, V., & Tsao, J. Y. (2018). Nurturing transformative US energy research: Two guiding principles. *MRS Energy and Sustainability, 5,* E10.

Natelson, D. (2018). Commentary: Condensed matter's image problem. *Physics Today, Vol. 71, Issue 12.* DOI:10.1063/pt.6.3.20181219a. https://physicstoday .scitation.org/do/10.1063/PT.6.3.20181219a/full/.

National Academies of Sciences, Engineering, and Medicine. (2019). *Frontiers of materials research: A decadal survey.* The National Academies Press.

National Research Council. (2014). Convergence: Facilitating transdisciplinary integration of life sciences, physical science, engineering, and beyond.

Nelson, R. R. (1959). The simple economics of basic scientific research. *Journal of Political Economy, 67*(3), 297–306.

Niaz, M. (2000). The oil drop experiment: A rational reconstruction of the Millikan–Ehrenhaft controversy and its implications for chemistry textbooks. *Journal of Research in Science Teaching, 37*(5), 480–508.

Nordhaus, W. D. (1996). Do real-output and real-wage measures capture reality? The history of lighting suggests not. In T. F. Bresnahan & R. J. Gordon (Eds.), *The economics of new goods* (pp. 27–70). University of Chicago Press.

Odumosu, T., Tsao, J. Y., & Narayanamurti, V. (2015). Commentary: The social science of creativity and research practice: Physical scientists, take notice. *Physics Today, 68*(11), 8–9.

OECD. (2015). Frascati manual 2015: Guidelines for collecting and reporting data on research and experimental development. https://www.oecd.org /publications/frascati-manual-2015-9789264239012-en.htm.

Ott, W. (2016, August 25). Katharine Gebbie's leadership and lasting impact. NIST Physical Measurement Laboratory. https://www.nist.gov/pml /katharine-gebbies-leadership-and-lasting-impact.

Pearl, J., & Mackenzie, D. (2018). *The book of why: The new science of cause and effect.* Basic Books.

Perrin, Jean (1970). *Les atomes.* Gallimard (original edition published in 1913).

Pielke, R. A., Jr. (2012). Basic research as a political symbol. *Minerva, 50*(3), 339–361.

Pierce, J. R. (1975). *Mervin Joe Kelly: A biographical memoir*. National Academy of Sciences.

Pigliucci, M. (2008). Is evolvability evolvable? *Nature Reviews Genetics, 9*(1), 75.

Planck, M. (1950). *Scientific autobiography: And other papers*. Open Road Media (2014 edition).

Plévert, L. (2011). *Pierre-Gilles de Gennes: A life in science*. World Scientific.

Polya, G. (2014). *How to solve it: A new aspect of mathematical method* (2nd ed.). Princeton University Press.

Popper, K. (2005). *The logic of scientific discovery*. Routledge.

Price, D. D. (1984). The science / technology relationship, the craft of experimental science, and policy for the improvement of high technology innovation. *Research Policy, 13*(1), 3–20.

Price, D. J. (1986). *Little science, big science . . . and beyond*. Columbia University Press.

Riordan, M., & Hoddeson, L. (1998). *Crystal fire: The invention of the transistor and the birth of the information age*. WW Norton and Company.

Rosenberg, N. (1974). Science, invention and economic growth. *The Economic Journal, 84*(333), 90–108.

Rosenberg, N. (1982). *Inside the black box: Technology and economics*. Cambridge University Press.

Rubin, G. M. (2006). Janelia Farm: An experiment in scientific culture. *Cell, 125*(2), 209–212.

Rybaczyk, P. (2005). *Expert network time protocol: An experience in time with NTP*. Apress.

Saltzer, J. H., Reed, D. P., & Clark, D. D. (1984). End-to-end arguments in system design. *ACM Transactions on Computer Systems, 2*(4), 277–288.

Sargent, J. F., Jr. (2018). Defense science and technology funding. *Congressional Research Service, R45110*.

Schawlow, A. L., & Townes, C. H. (1958). Infrared and optical masers. *Physical Review, 112*(6), 1940.

Schivelbusch, W. (1995). *Disenchanted night: The industrialization of light in the nineteenth century*. University of California Press.

Schmookler, J. (1966). *Invention and economic growth*. Harvard University Press.

Schumpeter, J. A. (1942). *Capitalism, socialism and democracy*. Harper and Brothers.

Scovil, H., Feher, G., & Seidel, H. (1957). Operation of a solid state maser. *Physical Review, 105*(2), 762.

Shakespeare, W. (2017). *Hamlet*. Folger Digital Library (original edition published 1603).

Shneiderman, B. (2016). *The new ABCs of research*. Oxford University Press.

Shockley, W. (1950). *Electrons and holes in semiconductors: With applications to transistor electronics*. van Nostrand.

Shockley, W. (1956). Transistor technology evokes new physics. *Nobel Lectures, Physics 1942–1962*, Elsevier Publishing Company, Amsterdam (1964).

Simon, H. A. (1962). The architecture of complexity. *Proceedings of the American Philosophical Society, 106*(6), 467–482.

Simon, H. A. (1977). *Models of discovery: And other topics in the methods of science.* Reidel.

Simon, H. A. (2001). Science seeks parsimony, not simplicity: Searching for pattern in phenomena. In A. Zellner, H. A. Keuzenkamp, & M. McAleer (Eds.), *Simplicity, inference and modelling: Keeping it sophisticatedly simple* (pp. 32–72). Cambridge University Press.

Simonton, D. K. (2004). *Creativity in science: Chance, logic, genius and zeitgeist.* Cambridge University Press.

Simonton, D. K. (2018). Defining creativity: Don't we also need to define what is not creative? *The Journal of Creative Behavior, 52*(1), 80–90.

Stanley, K. O., & Lehman, J. (2015). *Why greatness cannot be planned: The myth of the objective.* Springer.

Stephan, P. E. (1996). The economics of science. *Journal of Economic literature, 34*(3), 1199–1235.

Stokes, D. E. (2011). *Pasteur's quadrant: Basic science and technological innovation.* Brookings Institution Press.

Störmer, H., Gossard, A., Wiegmann, W., & Baldwin, K. (1981). Dependence of electron mobility in modulation-doped GaAs-(AlGa)As heterojunction interfaces on electron density and Al concentration. *Applied Physics Letters, 39*(11), 912–914.

Stormer, H. L. (1999). Nobel Lecture: The fractional quantum Hall effect. *Reviews of Modern Physics, 71*(4), 875.

Summers, L. H. (2002, August 1). Harvard University statement of values. https://www.harvard.edu/president/speech/2002/harvard-university-statement-values.

Thompson, D. (2017, November). Google X and the science of radical creativity. *The Atlantic.* https://www.theatlantic.com/magazine/archive/2017/11/x-google-moonshot-factory/540648/.

Townes, C. H. (1965). 1964 Nobel lecture: Production of coherent radiation by atoms and molecules. *IEEE Spectrum, 2*(8), 30–43.

Townes, C. H. (1999). *How the laser happened: Adventures of a scientist.* Oxford University Press.

Tsao, J., Boyack, K., Coltrin, M., Turnley, J., & Gauster, W. (2008). Galileo's stream: A framework for understanding knowledge production. *Research Policy, 37*(2), 330–352.

Tsao, J. Y., Han, J., Haitz, R. H., & Pattison, P. M. (2015). The blue LED Nobel Prize: Historical context, current scientific understanding, human benefit. *Annalen der Physik, 527*(5–6).

Tsao, J. Y., Ting, C. L., & Johnson, C. M. (2019). Creative outcome as implausible utility. *Review of General Psychology, 23,* 279.

Tsao, J. Y., & Waide, P. (2010). The world's appetite for light: Empirical data and trends spanning three centuries and six continents. *Leukos, 6*(4), 259–281.

Tsui, D. C., Stormer, H. L., & Gossard, A. C. (1982). Two-dimensional magneto-transport in the extreme quantum limit. *Physical Review Letters, 48*(22), 1559.

Uzzi, B., Mukherjee, S., Stringer, M., & Jones, B. (2013). Atypical combinations and scientific impact. *Science, 342*(6157), 468–472.

Weinberg, A. M. (1961). Impact of large-scale science on the United States. *Science, 134*(3473), 161–164.

Weisbuch, C. (2018). Historical perspective on the physics of artificial lighting. *Comptes Rendus Physique, 19*(3), 89–112.

Wikipedia contributors. (2019, March 1). A / B testing. Wikipedia. https://en .wikipedia.org/wiki/A/B_testing.

Williams, R. (2008). How we found the missing memristor. *IEEE Spectrum,* 28–35.

ACKNOWLEDGMENTS

Although we did not know it when we first began, the process of creating this book bears all the hallmarks of the nature and nurturing of research around which the book revolves. The book is as much about the science as it is about the engineering of research and research organizations. Both the questions we sought to answer and the answers we found to those questions morphed countless times in an intricate dance. Many of the book's insights surprised even us—our own punctuated equilibria of knowledge evolution.

The process of creating this book was far from smooth. Serious research and writing began in the fall of 2016 and culminated in spring 2021. We found many points of agreement born of common experiences in research policy, management, and practice, but more interesting were the differences, many born of Venky leaning toward holistic thinking and Jeff leaning toward reductionist thinking. The result was intense debates, many challenging to navigate, but that, in the end, enormously improved the final product. And while we know the book yet contains inconsistencies and loose ends, we hope it will benefit the institutions with which we are associated and, even more so, that it will provide broader public benefit to the world of research we both hold dear.

In many ways, this book reflects the culmination of our two careers: Venky at Bell Labs, Sandia National Laboratories the University of California–Santa Barbara, and Harvard—both the Paulson School of Engineering and Applied Science and the Harvard Kennedy School; and Jeff largely at Sandia National Laboratories but with brief forays into teaching, start-up companies, and advisory roles in support of the Department of Energy Office of Basic Energy Sciences and the Office of Energy Efficiency and Renewable Energy Solid-State Lighting Program. For both of us, 2008 was a pivotal year: Venky shifted his professional focus from university administration and personal research in condensed matter and materials physics to teaching and research at the intersections of science, technology, and public policy through a joint faculty appointment between Harvard's Paulson School of Engineering and Applied Sciences and Harvard's Kennedy School; and Jeff made his initial foray into the science of science with his "Galileo's Stream" collaboration with Kevin Boyack, Mike Coltrin, Wil Gauster, and Jessica Turnley. Subsequently, in 2013, Venky and Jeff teamed up with Greg Feist (San Jose State University) to co-organize an "Art and Science of Science and Technology" workshop at Sandia; in 2016, Venky and Tolu Odumosu researched and wrote a book on cycles of invention and discovery, while Jeff began his Belfer Fellowship at the Harvard Kennedy School in 2016.

We are grateful to many friends and colleagues along the way whose encouragement paved the way for this book.

Venky is particularly grateful to Graham Allison, who recruited him to succeed John Holdren as head of the Science, Technology Public Policy Program at the Belfer Center of Science and International Affairs at Harvard's Kennedy School in 2009, bequeathing him a group of talented young scholars working in the area of energy technology innovation; to Sheila Jasanoff, who was a very early faculty partner in teaching and who brought to his attention younger scholars in the field of science and technology studies and the literature of that field; and to his continuing collaborations with Tolu Odumosu (now at the University of Virginia), Laura Diaz Anadon (now at the University of Cambridge), and Ambuj Sagar (now at the Indian Institute of Technology, Delhi). Venky has also benefited greatly from his recent service on several panels, including "Advancing Research in Science and Engineering" and "Restoring the Foundation" (American Academy of Arts and Sciences). His thinking on science and technology policy has been influenced by many colleagues on those panels, particularly Nancy Andrews, Dan Mote, Cherry Murray, and Neal Lane.

Jeff is particularly grateful to colleagues who, at key junctures of a less-than-easy transition, were valued collaborators or moral supporters: Sarah Allendorf, Glory (Emmanuel) Avina, Kevin Boyack, Tom Brennan, Mike Coltrin, Mike Descour, Greg Frye-Mason, Wil Gauster, Diane Gaylord, Bob Hwang, Curtis Johnson, Billie Kazmierczak, Rob Leland, Jeff Nelson, Julie Phillips, Steve Rottler, Rick Schneider, Andy Schwartz, Susan Seestrom, Austin Silva, Jerry Simmons, Rickson Sun, Christina Ting, Jessica Turnley, Karen Waldrip, and Rieko Yajima. He is also indebted to a few individuals who were special inspirations to him at various stages of his career and major influences on how he views research: Roland Haitz, Harriet Kung, Paul Peercy, Tom Picraux, and Eli Yablonovitch. Jeff is also deeply appreciative of his longtime affiliation with Sandia National Laboratories, a multimission laboratory managed and operated by National Technology and Engineering Solutions of Sandia, LLC, a wholly owned subsidiary of Honeywell International Inc., for the US Department of Energy's National Nuclear Security Administration under contract DE-NA0003525. Any subjective views or opinions that might be expressed in the book do not necessarily represent the views of the US Department of Energy or the United States government.

We both are grateful to those who commented on various sections and drafts of this book: Laura Diaz Anadon, Mike Coltrin, Mike Descour, Anna Goldstein, Bob Hwang, Curtis Johnson, Dan Krupka, Ambuj Sagar, Jerry Simmons, Mike Smith, Jessica Turnley, Claude Weisbuch, two wonderful anonymous reviewers, and the Harvard University Press faculty editorial board. We are grateful to Janice Audet, our Harvard University Press editor, who championed the book and provided her own valuable comments, and to Stephanie Vyce and Emeralde Jensen-Roberts of HUP for their help shepherding the book at various stages. We benefited from the excellent administrative support we have received from staff at the Belfer Center and School of Engineering and Applied Sciences, especially Karin van der Schaaf, Sarah Lefebvre, and the late Patricia McLaughlin. We benefited enormously from Andi M. Penner's careful editing, which not only made the book more readable, but also served as a welcome catalyst for deeper thinking.

Our view that researchers must have a secure base and home from which intellectual and emotional risk can be taken has been shaped by the secure base and home that our own families have provided us. For Venky, his wife, Jaya, has been his beacon and constant source of support for six decades, along with his three children (Arjun, Ranjini, and Krishna) and seven grandchildren. For Jeff, his wife, Sylvia; his children, Evan, Emil, and Eugene; and his parents, Ching and Matilda, have been his life's greatest joys. We thank you all for enabling this labor of love.

punctuated equilibria, 5, 10, 12, 109–111, 125, 143, 146, 216, 227
Purcell, Edward Mills, 59

quadrants, 170–177; Bohr's, 173–177; Pasteur's, 168, 172–177; Townes's, 173–174, 176–177; Edison's, 173–176; Whafnium, 173–174; Si-Properties, 173–174; Apollo 11, 174; Moore's, 174
quest for fundamental understanding, 172
question-and-answer pairs and finding, 3, 8–10, 65–108, 121–125, 129–133, 133–143, 150–155, 177–179, 211, 217

radar, 34, 50, 56, 58, 122, 190
radical or disruptive versus standard engineering, 123, 153, 170, 216, 217
Ramsey, Norman, 59
recombination of ideas, 10, 82, 84, 85, 88, 91–92, 94, 107
reductionism versus constructivism, 5, 26, 77–79, 86, 87, 177, 227; reputational credit, 18, 189, 191
reputational reward, 18, 158, 186, 195, 197, 200, 203
research: accountability, 5–6, 13, 148–149, 159, 164–165, 186, 192–199, 203, 207; applied, 2, 51, 155, 175, 177–179, 211–212, 215; basic, 2, 51, 155, 171, 175, 177, 211–212, 215; engineering, 2, 155, 161, 170–175, 211–213, 215; foundational, 175, 211–213; is a deeply human endeavor, 2, 13, 21, 63, 199, 207; leadership, 6, 12, 149–150, 158–159, 161–165, 167–169, 173, 187, 189–203, 224; outcome versus impact, 155, 168, 188, 198, 200, 217; pure, 51, 175, 217; use-inspired, 168, 172–175, 177; "6.1," "6.2," and "6.3" research and development, 169
research environments reflect human relationships and group spirit, 198, 207

Research Labs 1.0, 2.0, 3.0, 206–207
Restoring the Foundation, 228
Rockefeller University, 149
Roosevelt, Franklin D., 208
Rosenberg, Nathan, 22, 95, 131
Rubin, Gerald, 154, 167

Ś, 16–17, 23–24, 26, 39, 49, 52, 55, 63, 81–82, 110–112, 114, 167–172, 212, 217
scaffolding, 116–117
Schawlow, Art, 60, 62, 64
Schumpeter, Joseph, 6, 11, 30, 126
science: big, 36, 40, 151, 161; information sciences and software, 14, 204; life sciences and medicine, 14, 164, 204; normal, 27, 37, 170, 216–217; physical sciences and engineering, 6, 14, 19, 44, 115, 164, 166, 204; revolutionary, 170; scientific hypothesis, 3, 36–38, 181
scientific method, 7–8, 26–29, 34–49, 95, 106, 121–123, 179–182, 217; explanation-finding, 40–46, 213; fact-finding, 34–40, 213; generalizing, 46–49, 214
Scovil, Henry, 61
seamless web of knowledge, 5, 47, 66, 67, 75, 97, 100, 109, 111, 123, 129, 147, 204
seek what you don't know, not what you know, 87, 197
Shakespeare, William, 23
Shockley, William, 10, 51–52, 55
Simon, Herbert, 6, 8, 19–20, 37, 71–73, 75, 180
Simonton, Dean, 85, 120
"singleton" versus "multiple" discoveries and inventions, 164
Si-Properties quadrant, 173–174
solution looking for a problem, 88, 102, 173
Sony, 34
specialist, 71, 92

special relativity, theory of, 1, 9, 11, 13,
17, 24, 31, 32, 41, 46, 66, 73, 94,
96–100, 105, 107, 114, 121, 126, 151,
169, 204
speed of light, 9, 11, 32, 36, 46, 96–100,
107, 112, 147, 183
spillover, knowledge, 33, 44, 139,
153–155, 176, 215
spot of Arago, 48
standard versus disruptive or radical
engineering, 170
stand on the shoulders of, 17, 49, 64,
82, 147, 153
Stanley, Ken, 125, 176
Stephan, Paula, 153
Stokes, Donald, 6, 23, 171–175
Stormer, Horst, 171–172, 182
stylized fact, 5, 7, 8, 10, 12, 15, 43, 63,
65–66, 106, 110, 130, 180, 217
Summers, Larry, 154, 159
supply push, 87
surprise versus consolidation.
See learning by surprise versus
consolidation

$\dot{T}$, 16–17, 23–24, 26, 39, 49, 55, 63,
81–82, 110–112, 114, 167–172,
212, 217
tacit, 9, 19, 33, 81, 114–115
technical breadth and depth, 90–93,
163, 186–189, 196–197
technology readiness levels, 169
technoscience, 5–6, 35, 49, 175, 177,
205, 218
technoscientific method, 7–8, 26–49,
49–63, 121–123, 167–185, 218. *See
also* scientific method; engineering
method
Thiel, Peter, 184
3M, 85
threshold effects, 11, 125, 130,
136–137, 140
tool for, versus object of,
observation, 36

Townes, Charles, 56–62, 64, 85, 89,
173–177
Townes's quadrant, 173–174, 176–177
transactional, 159, 161, 205
transistor, 1, 3, 4, 13, 17, 18, 32,
33, 46, 49, 50–56, 57, 64, 82–83,
91, 122, 129, 133, 144, 154,
172, 176, 204
transistor effect, 1, 3, 52, 55, 64,
144, 204
transportability, 47. *See also*
generalizing
Tsao, Jeffrey, 2, 4, 11, 25, 32, 121, 134,
141, 142, 147, 156, 159, 169
T-shaped generalist / specialist, 92, 217
Twain, Mark, 120

ultimate versus proximate goals, 25, 30,
68, 150–151. *See also* goal
uncertainty and certainty in research
and development, 26, 33, 45, 60, 139,
147, 158, 180, 185, 191
University of California at Santa
Barbara, 166
utility: curiosity versus practical
application, 21–24, 46, 168,
172–177, 188, 218; implausible
versus plausible, 118–121, 136, 144,
182–183, 185, 214; proximate versus
ultimate, 24–25; with learning as
cultural selection pressure on $\dot{S}$ and
$\dot{T}$, 23–25, 29–30, 35, 112, 128,
150–158, 194

vacuum tubes, 33, 50, 52, 122, 128
verification, 41, 44–46, 64, 180–181;
finding, 45; of discovery, 45;
seeking, 45
virtual organization, 161–162
Vrba, Elisabeth, 33

Wallace, Alfred, 85
Wall Street perspective on the primacy
of short-term and private return